Recombinant DNA Products:
Insulin, Interferon and Growth Hormone

Editor

Arthur P. Bollon, Ph.D.

Director of Genetic Engineering
Chairman, Department of Molecular Genetics
Wadley Institutes of Molecular Medicine
Dallas, Texas

CRC Press
Taylor & Francis Group
Boca Raton London New York

CRC Press is an imprint of the
Taylor & Francis Group, an **informa** business

First published 1984 by CRC Press
Taylor & Francis Group
6000 Broken Sound Parkway NW, Suite 300
Boca Raton, FL 33487-2742

Reissued 2018 by CRC Press

Library of Congress Cataloging in Publication Data
Main entry under title:

Recombinant DNA products.

Bibliography: p.
Includes index.
1. Recombinant DNA. 2. Genetic engineering.
3. Biological products. I. Bollon, Arthur P., 1942-
II. Title: Recombinant D.N.A. products. [DNLM: 1. DNA,
Recombinant. 2. Insulin--Biosynthesis. 3. Interferons--
Biosynthesis. 4. Somatotropin--Biosynthesis.
QU 58 R3124]
QH442.R383 1984   574.87'3282   84-7599
ISBN 0-8493-5542-2

A Library of Congress record exists under LC control number: 84007599

ISBN 13: 978-1-315-89715-8 (hbk)
ISBN 13: 978-1-351-07625-8 (ebk)

Visit the Taylor & Francis Web site at http://www.taylorandfrancis.com and the
CRC Press Web site at http://www.crcpress.com

PREFACE

The ability to isolate human genes, insert them into a microorganism which then produces a human protein, thereby serving as a biological factory, has already revolutionized biomedical research and purports to be the foundation for a new, highly lucrative industry. Of course, genetic engineering is not new, the novelty of recombinant DNA technology is the precision and efficiency with which scientists can manipulate genes. This book reviews advances made in recombinant DNA technology as it relates to the techniques employed, and the production and testing of potentially important products such as human interferon, insulin, and growth hormone.

The pioneering work of Paul Berg, Herbert Boyer, and Stanley Cohen, approximately 12 years ago, set the foundation for the present technology and represents one of the best testaments to the value of basic research. A prediction then that the analysis of bacterial restriction-modification would revolutionize cancer research, which at that time was receiving considerable priority, would have been considered at best a biased optimism on the part of the individuals working in that area. It is only in retrospect that certain decisions dominate. In prospect, such decisions are as influenced by temperament as by any scientific logistics. One obvious lesson to be learned from the development of recombinant DNA technology is that any attempt to minimize the commitment to basic studies will probably confer an enormous economic cost on our society.

Arthur P. Bollon, Ph.D.
Dallas, Texas
August 1, 1983

THE EDITOR

Arthur P. Bollon, Ph.D., is Director of Genetic Engineering and Chairman of the Department of Molecular Genetics at the Wadley Institutes of Molecular Medicine in Dallas, Texas.

Dr. Bollon received his B.A. degree in Biology from C.W. Post College in 1965, followed by a Fellowship to work on Microbial Genetics with Dr. Milaslav Demerec. In 1970, he received his Ph.D. in Microbiology at the Waksman Institutes of Microbiology at Rutgers University working on bacterial gene expression under Dr. Henry J. Vogel. Following a postdoctoral fellowship working on yeast genetics in the Department of Microbiology at Yale University, he joined the faculty of The University of Texas Health Science Center at Dallas in the Department of Biochemistry. Dr. Bollon presently maintains two adjunct faculty positions at North Texas State University and The University of Texas Health Science Center.

CONTRIBUTORS

Stanley Barban, Ph.D.
Scientist Administrator
Office of Recombinant DNA Activities
National Institute of Allergy and
 Infectious Diseases
National Institutes of Health
Bethesda, Maryland

Emily A. Barron, M.S.
Research Assistant
Department of Molecular Genetics
Wadley Institutes of Molecular Medicine
Dallas, Texas

John Baxter, M.D.
Professor of Medicine, Biochemistry and
 Biophysics
Metabolic Research Unit
University of California
San Francisco, California

Rama M. Belagaje, Ph.D.
Senior Biologist
Lilly Research Laboratories
Indianapolis, Indiana

Susan Berent, Ph.D.
Research Fellow
Department of Molecular Genetics
Wadley Institutes of Molecular Medicine
Dallas, Texas

Leon L. Bernhardt, M.D.
Senior Research Physician
Department of Medical Oncology
 and Immunology
Hoffmann-La Roche, Inc.
Nutley, New Jersey

Arthur P. Bollon, Ph.D.
Director of Genetic Engineering
Chairman, Department of Molecular
 Genetics
Wadley Institutes of Molecular Medicine
Dallas, Texas

Herbert Boyer, Ph.D.
Professor
Department of Biochemistry and Biophysics
University of California
San Francisco, California

Paul W. Bragg, Ph.D.
Postdoctoral Fellow
Department of Molecular Genetics
Wadley Institutes of Molecular Medicine
Dallas, Texas

Christina Chen, M.S.
Research Associate
Department of Molecular Biology
Genentech, Inc.
San Francisco, California

Wanda deVlamick, B.S.
Director
Regulatory Affairs
Cetus Corporation
Emeryville, California

David Dixon, M.S.
Research Assistant
Department of Molecular Genetics
Wadley Institutes of Molecular Medicine
Dallas, Texas

Zofia E. Dziewanowska, M.D., Ph.D.
Director
Department of Medical Oncology and
 Immunology
Hoffmann-La Roche, Inc.
Nutley, New Jersey

Seymour Fein, M.D.
Senior Research Physician
Department of Medical Oncology
 and Immunology
Hoffmann-La Roche, Inc.
Nutley, New Jersey

Motohiro Fuke, Ph.D.
Senior Scientist
Department of Molecular Genetics
Wadley Institutes of Molecular Medicine
Dallas, Texas

Frank Hagie, B.S.
Research Assistant
Molecular Biology Department
Genentech, Inc.
San Francisco, California

Cheryl Hendrix, B.S.
Research Assistant
Department of Molecular Genetics
Wadley Institutes of Molecular Medicine
Dallas, Texas

Norwood O. Hill, M.D.
President
Wadley Institutes of Molecular Medicine
Dallas, Texas

Raymond L. Hintz, M.D.
Associate Professor
Department of Pediatrics
and
Head
Division of Pediatric Endocrinology
Stanford University Medical Center
Stanford, California

Ronald A. Hitzeman, Ph.D.
Scientist
Molecular Biology Department
Genentech, Inc.
San Francisco, California

Hansen M. Hsiung, Ph.D.
Senior Biologist
Lilly Research Laboratories
Indianapolis, Indiana

Keiichi Itakura, Ph.D.
Director
Molecular Genetics Department
The Beckman Research Institute of the
 City of Hope
Duarte, California

June M. Lugovoy, B.S.
Research Assistant
Genentech, Inc.
San Francisco, California

Massoud Mahmoudi, B.S.
Research Assistant
Department of Molecular Genetics
Wadley Institutes of Molecular Medicine
Dallas, Texas

Nancy Mayne, B.S.
Biologist
Lilly Research Laboratories
Indianapolis, Indiana

W. Courtney McGregor, Ph.D.
Assistant Director
Biopolymer Research Department
Hoffmann-La Roche, Inc.
Nutley, New Jersey

Elizabeth A. Milewski, Ph.D.
Scientist Administrator
Office of Recombinant DNA Activities
National Institute of Allergy and Infectious
 Diseases
National Institutes of Health
Bethesda, Maryland

Armin H. Ramel, Ph.D.
Process Research and Development
 Department
Genentech, Inc.
San Francisco, California

Philip Raskin, M.D.
Associate Professor
Department of Internal Medicine
University of Texas Health Science Center
Southwestern Medical School
Dallas, Texas

Arthur D. Riggs, Ph.D.
Chairman
Division of Biology
The Beckman Research Institute of the
 City of Hope
Duarte, California

Rajinder Sidhu, Ph.D.
Research Fellow
Department of Molecular Genetics
Wadley Institutes of Molecular Medicine
Dallas, Texas

Arjun Singh, Ph.D.
Scientist
Department of Molecular Biology
Genentech, Inc.
San Francisco, California

Nowell Stebbing, Ph.D.
Vice President for Scientific Affairs
Amgen
Thousand Oaks, California

Richard Torczynski, M.S.
Research Assistant
Department of Molecular Genetics
Wadley Institutes of Molecular Medicine
Dallas, Texas

Phillip K. Weck, Ph.D.
Clinical Research Scientist
Immunology Section
Department of Clinical Investigation
Burroughs Wellcome Company
Research Triangle Park, North Carolina

TABLE OF CONTENTS

Chapter 1

RECOMBINANT DNA TECHNIQUES: ISOLATION, CLONING, AND EXPRESSION OF GENES

A. P. Bollon, E. A. Barron, S. L. Berent, P. W. Bragg, D. Dixon, M. Fuke, C. Hendrix, M. Mahmoudi, R. S. Sidhu, and R. M. Torczynski

TABLE OF CONTENTS

I. INTRODUCTION

The cloning and expression of foreign genes using recombinant DNA technology has permitted access to complex biological mechanisms such as eucaryotic RNA splicing, oncogene dynamics, and developmental systems such as antibody diversity. In addition, the technology has been the foundation for a new bio-technology industry. This chapter contains an analysis of some of the recombinant DNA techniques that have been employed for the manipulation of foreign genes in microorganisms resulting in protein production as diagrammed in Figure 1. Subsequent chapters will address the expression, clinical trials and production of genetically engineered human interferon, insulin, and growth hormone.

II. ENZYMES

The isolation and cloning of genes involves a series of linked enzymatic steps. Experience with two-enzyme coupled reactions is enough to sensitize the researcher to the complexity of linking five or more enzyme reactions, that are characteristic of the steps involved in the synthesis and cloning of cDNA. Considering that the substrates as well as the catalysts are biological macromolecules, it is not surprising that there are various opinions as to the most efficient protocols. The purity and correct storage procedures for the substrate and enzymes are clearly very critical. Selected restriction enzymes and other enzymes which are commonly used for cloning procedures will be discussed with emphasis on some of their characteristics and utility.

A. Restriction Endonucleases

Restriction endonucleases are enzymes which have been identified in prokaryotic organisms and recognize specific DNA sequences for their endonucleolytic activity. This mechanism permits organisms to prevent foreign DNA from integrating into their genome, which could jeopardize the genetic integrity of the species. Since bacterial sexuality involves direct movement of DNA between organisms by transformation, conjugation, or transduction, it is not surprising that some mechanism evolved to protect against undesirable DNA (*restriction*) as well as to protect native endogenous DNA (*modification*).

Three types of restriction endonucleases have been characterized as indicated in Table 1.

Type II enzymes have been most useful for DNA cloning due to the separation of the endonuclease and methylation activities into separate enzymes and the sequence specificity of the endonucleolytic action. A key feature of many restriction endonucleases is their asymmetric cleavage generating single-stranded ends. For example, the commonly used enzyme EcoRI recognizes and cleaves the following sequence at the arrows.

GENE MANIPULATION

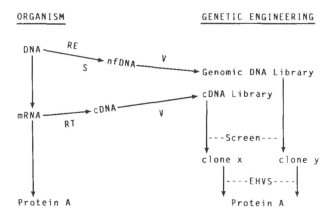

FIGURE 1. cDNA, copy DNA; EHVS, expression host-vector system; nf DNA, natural fragmented DNA; RE, restriction endonuclease; S, sheared; V, vector. Protein A is a protein made either naturally by the organism or by genetic engineering. Clone X is a single clone identified from a cDNA library containing Protein A cDNA. Clone Y is a single clone identified from a genomic library containing a Protein A natural gene.

Table 1
RESTRICTION ENZYME TYPES

Restriction Enzyme Characteristics	Restriction Enzyme Classes[a]		
	Type I	Type II	Type III
Enzyme contains *both* modification (methylase) and endonuclease activity	+	−	+
Enzyme contains only endonuclease activity and a separate enzyme is involved in methylation	−	+	−
Recognize unmethylated sequences for endonuclease cleavage	+	+	+
Sequence specific cleavage	−	+	+
Class most useful for DNA cloning	−	+	−

[a] An enzyme type exhibiting the specific characteristic is indicated by a +.

$$——\overset{5'\ \downarrow}{\text{GAATTC}}\overset{3'}{}——$$

$$——\underset{3'}{\text{CTTAAG}}\underset{\uparrow\ 5'}{}——$$

Such a cleavage generates two fragments containing single-stranded ends.

$$——\overset{5'}{\text{G}}\qquad \text{AATTC}\overset{3'}{}——$$
$$+$$
$$——\underset{3'}{\text{CTTAA}}\qquad \text{G}\underset{5'}{}——$$

Since the enzyme cuts between the GA and the recognition sequence is an inverted repeat, complementary single-stranded ends are generated. These fragments can either be religated or new DNA combinations can be generated by mixing DNA fragments containing complementary single-stranded tails. The restriction endonucleases can generate either single-stranded 5' ends such as above, 3' ends, or no single-stranded tails if the cleavage is at the center of the recognition sequence.

Although most of the restriction enzymes recognize different sequences, there is sequence overlap for some enzymes. Such enzymes which recognize similar sequences are called *isoschizomers*. As can be seen in Table 2, several enzymes such as XmaI and SmaI are isoschizomers, and several enzymes share compatible cohesive ends.

The interplay of enzyme recognition sites and the length of the recognition sequence bear on the efficiency of cloning and retrieval of specific genes from vectors. For example, MboI digested DNA can be cloned into a vector cleaved with BamHI but would not be retrievable by BamHI digestion. Another example is the useful capability of cloning a BglII digested DNA fragment into a vector cut with BamHI, such as pBR322. In this case the generated sequence AGATCC is not cut by either BamHI or BglII. Hence, the use of different restriction enzymes for cleavage at homologous sites or for ligation of cleaved fragments should be tempered with anticipation of the resulting sequences generated.

For restriction enzyme analysis of a cloned DNA fragment, the status of the host bacterial methylation system is important. *E. coli* contains the *dam* and *dcm* methylase systems which have different specificities. For cleavage of DNA by BclI, MboI, ClaI, Xba, TaqI, MboII, and HphI, the DNA should be prepared from strains of *E. coli* which is *dam*⁻. For cleavage of DNA by HhaI, HpaII, MspI, SalI, and XhoI the DNA should be generated from strains of *E. coli* that are *dcm*⁻ (Roberts et al., 1980).[1]

B. Other Enzymes Used in Gene Isolation and Manipulation

1. DNA Polymerase I

E. coli DNA polymerase I is a 109,000 dalton protein containing three activities: (1) 5'→3' polymerase activity using a 3'-OH primer and a single-stranded template, (2) 5'→3' exonuclease activity which degrades double-stranded DNA at free 5' ends, (3) 3'→5' exonuclease activity which degrades ss DNA and ds DNA from 3'-OH ends. The 5'→3' polymerase activity blocks the 3'→5' exonuclease activity.

One of the major uses of the DNA polymerase I is *nick-translation* labeling of DNA fragments. This is possible due to the polymerase and exonuclease activities. The enzyme can remove bases from the 5' side of the nick of treated DNA due to its 5'→3' exonuclease activity and add labeled bases to the 3' side due to its polymerase activity (Kelly et al., 1970).[2] DNA probes can be prepared for hybridization with specific activities in excess of 10^8 cpm/μg (Maniatis et al., 1975).[3] The high specific activity depends on the specific activity of the labeled bases and incorporation of about 30% of the [α-³²P] dNTPs into the DNA being nick-translated (Maniatis et al., 1982).[4]

2. Klenow Fragment of E. coli DNA Polymerase I

The Klenow fragment is a peptide of 76,000 daltons generated by cleavage of DNA polymerase I with subtilisin (Jacobsen et al., 1974).[5] Enzymatic activities include the 5'→3' polymerase activity and the 3'→5' exonuclease activity of the DNA polymer but not the 5'→3' exonuclease activity.

This enzyme is used for sequencing DNA using the Sanger dideoxy system (Sanger et al., 1977),[6] filling the 3' recessed termini of restriction enzyme treated DNA, labeling the termini of DNA fragments and second-strand cDNA synthesis involved in cDNA procedures. Labeling of DNA termini containing 3' extensions is more efficient when T4 DNA polymerase is utilized (Maniatis et al., 1982).[4]

Table 2

RESTRICTION ENZYMES CHARACTERISTICS

Enzyme	Microorganism	Recognition Sequence	Common Isoschizomers[a]	Compatible Cohesive Sites	Number of Cleavage Sites		
					Lambda	SV40	pBR322
AluI	*Arthrobacter luteus*	AG↓CT		Blunt	>50	35	16
ApyI	*Arthrobacter pyridinolis*	CC↓(A/T)GG	AtuI, EcoRII		>35	16	6
AsuI	*Anabaena subcylindrica*	G↓GNCC	Sau961		>30	11	15
AsuII	*Anabaena subcylindrica*	TT↓CGAA	MlaI	AccI, AcyI, ClaI, HpaII, TaqI	7	0	0
AtuII	*Agrobacterium tumefaciens* ID 135	CC↓(A/T)GG	EcoRII		>35	16	6
AvaI	*Anabaena variabilis*	G↓PyCGPuG		SalI, XhoI, XmaI	8	0	1
AvaII	*Anabaena variabilis*	G↓G(A/T)CC	AflI, BamN$_x$I	Sau961	>17	6	8
AvrII	*Anabaena variabilis*	CCTAGG			2	2	0
BalI	*Brevibacterium albidum*	TGG↓CCA	BstI	Blunt	15	0	1
BamHI	*Bacillus amyloliquefaciens* H	G↓GATCC	BstI	BclI, BglII, MboI, Sau3A, XhoII	5	1	1
BbvI	*Bacillus brevis*	GC(↓N)GC			>30	23	21
BclI	*Bacillus caldolyticus*	T↓GATCA		BamHI, BGlII, MboI, Sau3A, XhoII	7	1	0
BglI	*Bacillus globigii*	GCCNNNN↓NGGC			22	1	3
BglII	*Bacillus globigii*	A↓GATCT		BamHI, BclI, MboI, Sau3A, XhoII	6	0	0
BstEII	*Bacillus stearothermophilus* ET	G↓GTNACC	AspAI, BsPI		11	0	0
BstNI	*Bacillus stearothermophilus*	CC↓(A/T)GG	EcoRII		>35	16	6
ClaI	*Caryophanon latum* L.	AT↓CGAT		AccI, AcyI, AsyII, HpaII, TaqI	15	0	1
DdeI	*Beaulfovibrio desulfuricana* Norway strain	C↓TNAG			>50	19	8
DpnI	*Diplococcus pneumoniae*	G^{Me}A↓TC	Sau3A	Blunt	only cleaves methylated DNA		
EcoRI	*Escherichia coli* RY13	G↓AATTC			5	1	1
EcoB	*Escherichia coli* B	TGANN8			Type 1	1	0

Table 2 (continued)
RESTRICTION ENZYMES CHARACTERISTICS

Enzyme	Microorganism	Recognition Sequence	Common Isoschizomers[a]	Compatible Cohesive Sites	Lambda	SV40	pBR322
EcoK	*Escherichia coli* K	AACN6			Type I	Type I	2
EcoPI	*Escherichia coli* (PI)	AGACC			Type III	Type III	4
EcoRI'	*Escherichia coli* RY13	(A/G)A ↓ T(T/C)		Blunt	>10	24	15
EcoRII	*Escherichia coli* R245	↓CC(A/T)GG	AtuI, ApyI		>35	16	6
Fnu4HI	*Fusobacterium nucleatum* 4H	GC ↓ NGC		Blunt	>50	25	42
FnuDII	*Fusobacterium nucleatum* D	CG ↓ CG	ThaI	Blunt	>50	0	23
HaeI	*Haemophilus aegyptius*	(A/T)GG ↓ CC(A/T)		Blunt	?	11	7
HaeII	*Haemophilus aegyptius*	PuGCGC ↓ Py			>30	1	11
HaeIII	*Haemophilus aegyptius*	GG ↓ CC	BspRI, BsuRI	Blunt	>50	19	22
HgaI	*Haemophilus gallinarum*	GACGCN5 ↓ CTGCGCN10			>50	0	11
HhaI	*Haemophilus haemolyticus*	GCG ↓ C	FnuDIII, HinP,I		>50	2	31
HincII	*Haemophilus influenzae* R_c	GTPy ↓ PuAC	HindII	Blunt	34	7	2
HindII	*Haemophilus influenzae* R_d	GTPy ↓ PuAC	HincII, HinJCI	Blunt	34	7	2
HindIII	*Haemophilus influenzae*	A ↓ AGCTT	HsuI		6	6	1
HinfI	*Haemophilus influenzae* R_f	G ↓ ANTC	FnuAI		>50	10	10
HpaI	*Haemophilus parainfluenzae*	GTT ↓ AAC		Blunt	13	4	0
HpaII	*Haemophilus parainfluenzae*	C ↓ CGG	HapII, MnoI	AccI, AcyI, AsuIIClaI, TaqI	>50	1	26
HphI	*Haemophilus parahaemolyticus*	GGTGAN6 ↓ CCACTN7			>50	4	12
KpnI	*Klebsiella pneumoniae* OK8	GGTAC ↓ C		BamHI, BclI, BglII, XhoII	2	1	0
MboI	*Moraxella bovis*	↓ GATC	DpnI, Sau3AI		>50	8	22
MboII	*Moraxella bovis*	GAAGAN8 ↓ CTTCTN7			>50	16	11
MnlI	*Moraxella nonliquefaciens*	CCTC			>50	51	26
MspI	*Moraxella species*	C ↓ CGG	HpaII		>50	1	26
MstI	*Microcoleus species*	TGCGCA	AosI, FdiII		>10	0	4
PstI	*Providencia stuartii* 164	CTCGA ↓ G	SalPI, SflI		18	2	1

Enzyme	Organism	Sequence	Isoschizomers	Ends/Notes			
PvuI	*Proteus vulgaris*	CGATCG	NbiI, RshI		3	0	1
PvuII	*Proteus vulgaris*	CAG↓CTG		Blunt	15	3	1
SacI	*Streptomyces achromogenes*	GAGCT↓C	SstI		2	0	0
SacII	*Streptomyces achromogenes*	CCGC↓GG	CscI, SstII		>25	0	0
SacIII	*Streptomyces achromogenes*	ACGT			>100	?	?
SalI	*Streptomyces albus subspecies pathocidicus*	G↓TCGAC	HgiCIII, HgiDII	AvaI, XhoI	1	?	0
Sau3A	*Staphylococcus aureus* 3A	↓GATC	MboI	BamHI, BclI, BglII, MboI, XhoII	>50	8	22
Sau961	*Staphylococcus aureus* PS96	G↓GNCC	AsuI		>30	11	15
SmaI	*Serratia marcescens* S_b	CCC↓GGG	XmaI	Blunt	3	0	0
SphI	*Streptomyces phaeochromogenes*	GCATG↓C			4	2	1
SstI	*Streptomyces* Stanford	GAGCT↓C	SacI		2	0	0
SstII	*Streptomyces* Stanford	CCGC↓GG	SacII		4	0	0
SstIII	*Streptomyces* Stanford	ACGT	SacIII		>100	?	?
TaqI	*Thermus aquaticus* YTI	T↓CGA		AccI, AcyI, AsuII, ClaI, HpaII	>50	1	7
ThaI	*Thermoplasma acidophilum*	CG↓CG	FnuDII	Blunt	>50	0	23
XbaI	*Xanthomonas badrii*	T↓CTAGA			1	0	0
XhoI	*Xanthomonas holcicola*	C↓TCGAG	BluI, PaeR7I	AvaI, SalI	1	0	0
XhoII	*Xanthomonas holcicola*	(A/G)↓GATC(C/T)		BamHI, BclI, BglII, MboI, Sau3A	>20	3	8
XmaI	*Xanthomonas malvacearum*	C↓CCGGG	SmaI	AvaI	3	0	0
XmaIII	*Xanthomonas malvacearum*	C↓GGCCG			2	0	1

[a] Presented are representative isoschizomers. A complete list can be obtained in the New England Biolabs' catalogue, 1982—1983.

3. T4 DNA Polymerase

The T4 DNA polymerase also contains a $5' \rightarrow 3'$ polymerase activity and a very active $3' \rightarrow 5'$ exonuclease activity. Blunt-ended DNA can be labeled by exchange reaction where only one dNTP is present in the reaction and the $3' \rightarrow 5'$ exonuclease activity will degrade double-stranded DNA from the 3'-OH end until it exposes a base complementary to the dNTP included in the reaction. Other uses involve the labeling of DNA fragments for use as hybridization probes, labeling protruding 5' ends of DNA fragments by treatment with $3' \rightarrow 5'$ exonuclease and filling in the displaced bases with ^{32}P dNTPs. This replacement protocol can generate more extensive and selective labeling than nick-translation (Maniatis et al., 1982).[4]

4. Terminal Deoxynucleotidyl Transferase

The terminal transferase isolated from calf thymus catalyzes the addition of dNTP to the 3'-OH of DNA molecules, (Bollum, 1974).[7] The templates are either single-stranded DNA or double-stranded DNA with single-stranded 3'-OH ends. If Co^{++} is included in the reaction, blunt-ended ds DNA can be used as template.

One of the primary uses of the terminal transferase is the tailing of vectors and cDNA with complementary bases thus permitting the cloning of the cDNA fragments. In addition, the terminal transferase can be used for labeling of 3' ends of DNA fragments (Tu and Cohen, 1980).[8]

5. T4 Polynucleotide Kinase

The T4 polynucleotide kinase is isolated from T4 infected *E. coli* and catalyzes the transfer of γ-phosphate of ATP to a 5'-OH end in DNA or RNA (Richardson, 1971).[9] In the presence of excess ADP the kinase will also catalyze an exchange reaction where P^{32} is transferred from ATP to a single- or double-stranded DNA with a 5'-P terminus.

Uses for the kinase involve the labeling of 5' termini of DNA for Maxam and Gilbert DNA sequencing (Maxam and Gilbert, 1977)[10] and the phosphorylation of DNA lacking 5'-P termini.

6. Reverse Transcriptase

The RNA dependent DNA polymerase, or reverse transcriptase, utilized for most studies is an enzyme which is coded for by avian myeloblastosis virus. It contains two peptides, one of which catalyzes the $5' \rightarrow 3'$ polymerase activity using RNA or DNA as template, and a $5' \rightarrow 3'$ riboexonuclease activity using RNA-DNA hybrids as substrate (Verma, 1977).[11]

Reverse transcriptase is one of the most important enzymes in molecular biology since it catalyzes the synthesis of cDNA from an RNA template. In addition, the reverse transcriptase can be used for the labeling of termini of DNA with extended 5' ends.

The use of reverse transcriptase for cDNA synthesis requires very pure enzyme. Although several vendors sell the enzyme, different batches should be analyzed for contaminating RNase activity. Inhibitors of such activity, i.e., vanadyl-ribonucleoside complex, can also be added to the reaction mixture. Reports of contaminating DNase activity have also emphasized the importance of the purity of the enzyme if a reasonable yield of cDNA is to be generated.

As to the length of the cDNA, longer transcripts are generated with potassium compared to sodium ions and the enzyme is most efficient at 6 to 10 mM Mg^{++} for activity (Maniatis et al., 1982).[4]

7. T4 DNA Ligase

The T4 DNA ligase is a peptide of 68,000 daltons and catalyzes the formation of a phosphodiester bond between 3'-OH and 5'-P ends in DNA using double-stranded DNA molecules with cohesive ends as substrate. Blunt-end ligation also is possible and is enhanced by the presence of T4 RNA Ligase (Sugino et al., 1977).[12]

8. Exonuclease III

E. coli exonuclease III contains several activities. It has a $3' \rightarrow 5'$ exonuclease activity using ds DNA containing a 3'-OH end as template, an RNase-H activity and a 3' phosphatase activity which uses ds or ss DNA containing 3'-phosphate termini as substrate.

The exonuclease is used for generating linear template DNA for the dideoxy sequencing technique (Sanger et al., 1977)[6] and generating staggered ends on ds DNA due to its $3' \rightarrow 5'$ exonuclease activity.

9. λ Exonuclease

The λ exonuclease recognizes the terminal 5' phosphate of double-stranded DNA for its exonuclease activity (Little et al., 1967).[13] Its primary use is the removal of protruding 5' terminus from ds DNA which is needed for the terminal transferase tailing of DNA.

III. VECTORS

A plethora of specialized vectors have been created for gene cloning in *Escherichia coli*, *Saccharomyces cerevisiae* and various mammalian cells. The refinement of the vectors is exemplified by shuttle vectors which permit movement of vectors between bacteria-yeast or bacteria-animal cells. In fact, proprietary concerns related to specialized vectors are complex due to the overlap of information contained in the various vectors.

Some of the important properties of vectors which make them useful cloning vehicles are: (1) origins of replication for autogenous replication in host cells, (2) selective markers, (3) limited number of useful restriction enzyme sites, (4) high copy number, and (5) strong promoters if expression is desirable.

A. *E. coli* Plasmid Cloning Vectors

Useful plasmids for cloning genes in *E. coli* are *pBR322* and *pBR325* as can be seen in Table 3. Both plasmids contain genes which confer resistance to ampicillin (Ap^R) and tetracycline (Tc^R). pBR325 also contains a gene which confers resistance to chloramphenicol (Cam^R). Single restriction sites in these resistance genes permit insertion of foreign DNA and the subsequent selection for inactivation of the resistance gene as indicated in Table 3. Hence, the plasmid pBR322 when transformed into an *E. coli* which is Ap^S, Tc^S will become Ap^R, Tc^R. If foreign DNA is inserted into the Pst site in the Ap^R gene the transformed cells will be Ap^S, Tc^R. Inactivation selection for plasmids containing inserted DNA can be used for all three resistance genes using the appropriate plasmid and restriction enzyme.

B. *Saccharomyces cerevisiae* Cloning Vectors

Following the development of the techniques for transformation of yeast by Hinnen and Fink, 1978,[19] shuttle vectors for cloning genes in yeast and bacteria have been developed (Bollon and Silverstein, 1982).[20] They have been very useful for analysis of yeast gene organization and expression and for the production by yeast of foreign proteins such as human interferon. The key features of these vectors are their ability to replicate in either yeast or bacteria and the fact that they contain selective markers for both organisms.

As indicated in Table 4, there are basically three types of vectors for *S. cerevisiae*. YIp *vectors* contain markers for both yeast and bacteria and integrate at homologous regions of the yeast genome (Botstein and Davis, 1982).[21] These plasmids are extremely useful for studying gene organization and regulation in normal chromosomal microenvironments. In vitro modification of isolated genes and reinsertion back to homologous chromosomal po-

Table 3
COMMON *E. COLI* PLASMID CLONING VECTORS

Plasmids	Size (kb)[a]	Gene Markers	Single Restriction Site	Comments	Investigators	Ref.
Col E1	6.5	col, imm	EcoRI (CP), SmaI (CP)	Contains gene for colicin E1 which is induced by mitomycin C. Used for cloning and in construction of composite plasmids	Hershfield et al. (1974)	14
pBR322	4.3	Ap^R, Tc^R	AvaI, BamHI (Tc^R), ClaI, EcoRI, HindIII(Tc^R), Pst1(Ap^R), PvuI(Ap^R), PvuII, SalI(Tc^R)	Most commonly used plasmid vector for cloning in *E. coli*	Bolivar et al. (1977) Sutcliff (1979)	15 16
pBR325	5.7	Ap^R, Tc^R, Cm^R	AvaI, BamI(Tc^R), EcoRI(Cm^R), HindIII(Tc^R), PstI(Ap^R), PvuI(Ap^R), SalI(Tc^R)	Similar to pBR322 except one can do positive selection for inserts in the EcoRI site by Cm inactivation	Soberon et al. (1978)	17
pKC7	5.8	Ap^R, Kan^R	BamHI, BclI(Kan^R), BglII(Kan^R), BstEII, EcoRI, HindIII, PvuI(Ap^R), SmaI, XhoI	Kan^R inactivation selection by insertion at BglII and BclI and Ap^R inactivation selection by insertion at PstI	Rao and Rogers (1979)	18

[a]   Kilobases.

Table 4

COMMON PLASMID CLONING VECTORS FOR *SACCHAROMYCES CEREVISIAE*

Plasmids	Size (kb)[a]	Gene Markers	Single Restriction Site	Comments	Investigators	Ref.
YIp5	5.5	Ap^R, Tc^R, Ura3	BamHI(Tc^R), EcoRI, HindIII(Tc^R), SalI(Tc^R)	Integrate to homologue in yeast genome. Low copy number and poor yeast transformation efficiency. Useful for gene analysis.	Botstein and Davis (1982)	21
YEp13	10.7	Ap^R, Tc^R, Leu2	BamHI(T^R), HindIII(Tc^R), SalI(T^R)	Contain parts of yeast 2 micron plasmid. Transform yeast 10^3—10^4 times better than YIp. High copy number and stable.	Broach et al. (1979)	22
YRp17	6.9	Ap^R, Tc^R, Trp1, Ura3	BamHI(Tc^R), EcoRI, HindIII(Tc^R), SalI(Tc^R)	Contain yeast autonomous replicating sequences ARS and transform well but unstable.	Stinchcomb et al. (1979)	23
pSB226	13.5	Ap^R, Tc^R, Trp1, Leu2	BamHI(T^R), SalI(T^R)	Autonomous replication in yeast. Two yeast and two bacterial markers. High copy number.	Barron et al. (1983)	24

[a] Kilobases.

sitions permit the delineation of various control regions such as promoters. Homologous recombination of transformed genes is unique to yeast and has not been accomplished for most animal cell cloning systems. This property of the yeast system represents a major advantage for studying directed gene control studies.

The *YEp vector* is another class of yeast vectors that contains part or all of the yeast 2-μm DNA as well as bacterial plasmid replicating origins. The YEp vectors have a higher copy number (25 to 75 copies) than YIp vectors due to the presence of the 2-μm DNA. The yeast plasmid 2-μm DNA exists at about 50 copies per cell. Another advantage of YEp vectors is their higher yeast transformation efficiency.

As indicated in Table 4, the *YRp vectors* contain autonomous replicating sequences (ARS), which permit multiple copies although lower and less stable than the YEp vectors. The ARS sequences have been isolated from yeast nuclear genes such as TrpI (Stinchcomb et al., 1979).[23] The YRp vector does integrate into the homologous region of the genome like YIp. Due to their higher copy number than YIp vectors, YRp vectors could be most useful for generation of gene integrates for analysis. The instability of YRp vectors can be rectified by incorporating sequences from the yeast centromere (Clarke and Carbon, 1980).[25]

We have constructed *vector pSB226*, which is related to the YEp vector class as indicated in Figure 2. It contains the whole 2-μm DNA and two yeast markers Trp1 and Leu2 (Barron et al., 1983).[24] The pSB226 copy number appears to be greater than YEp13 and should be useful for yeast minichromosome analysis and production of human proteins.

C. Specialized Vectors

Several vectors have been constructed for specialized functions. The series of Charon λ vectors generated by Blattner, 1977,[26] have been most useful for genomic cloning of complete or partially restricted DNA fragments. Each charon is unique in that they are capable of integrating inserts of a specific size range due to the bias of the λDNA packaging process. One example is λ 4A which can be utilized for the construction of libraries of 15 to 20 kb DNA fragments as indicated in Table 5. The use of these specialized vectors for creating genomic libraries will be discussed below.

Single-strand viral vectors such as *M13 mp*, Table 4, have become very useful for DNA sequencing (Sanger et al., 1977),[6] DNA probe generation and cloning of either DNA strand of restriction enzyme fragments (Messing et al., 1982).[28] In addition, M13 mp7, which also contains a beta-galactosidase gene-linker system also present in M13 mp8, has been used for the production by *E. coli* of human alpha 2 interferon (Slocombe et al., 1982).[31]

Cosmids are vectors which are used for cloning large DNA fragments, 30 to 45 kb. Most bacterial plasmids containing such large inserts would have greatly reduced transformation efficiencies and the optimal size inserts for lambda vectors are about 20 kb (Hohn and Hinnen, 1980).[29] Cosmid vectors contain the usual plasmid markers and cohesive end sites (*cos*) from lambda phage (Collins and Hohn, 1978).[30] The *cos* sequences are recognized by lambda specific packaging proteins and lambda head-precursors that permit cosmids to be packaged into lambda-capsids. Large fragments of DNA (30 to 45 kb) can be cloned because lambda-capsids require DNA of a minimal size for packaging; the lambda genome size is 50 kb. As indicated in Table 5, cosmid pHc79 is only 6.4 kb and therefore would need a large insert for it to become packaged. Cosmids can be packaged if inserts placed between the *cos* sequences result in their separation by about 37 to 52 kb. Therefore, cosmids can be used to clone large DNA fragments and still retain a reasonable transformation frequency (10^5 transformants/mg DNA). Also indicated in Table 5 is a yeast cosmid, pYc1, which contains components of yeast plasmid YEp6 that permit the cosmid to replicate in yeast (Hohn and Hinnen, 1980).[29]

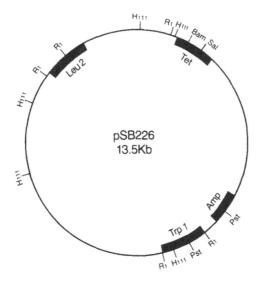

FIGURE 2. HIII is HindIII; RI is EcoRI; Tet is tetra-cycline resistance loci; AMP is ampicillin resistance loci; kb is kilobases.

IV. GENE CLONING

A. Synthesis of cDNA

As indicated in Figure 1, cDNA complementary to specific mRNA can be made by reverse transcriptase, the second DNA strand synthesized and both strands cloned in one of the vectors described in Tables 3 to 5 (Efstratiadis et al., 1976).[32] Most of the initial cDNA work involved standard vectors such as pBR322, vectors designed specifically for cDNA cloning will be discussed later. Some properties of the reverse transcriptase, and *E. coli* DNA polymerase I or Klenow fragment which can be used to synthesize the second cDNA strand have been described above. Since the first cDNA strand loops back and acts as a primer for the second strand, the hairpin loop must be cleaved by the single-strand specific nuclease S1. The integrity of the double-stranded cDNA following S1 nuclease digestion is a major determinant in the efficiency of cDNA synthesis and cloning. Since the ds cDNA termini may not be blunt ended, treatment with the Klenow fragment can repair any staggered ends due to the S1 nuclease digestion and can result in an improved cloning efficiency (Seeburg et al., 1977).[33]

The ds cDNA can be cloned by adding about 20 dG's to the cDNA with terminal transferase and the tailed cDNA can be ligated to a vector tailed with 20 dC's (Villa-Komaroff et al., 1978 and Peacock et al., 1981).[34,35] The advantage of dG-dC homopolymer-tail cloning is that the dG extension of vectors at the PstI site results in regeneration of the PstI cleavage site, permitting recovery of the inserted cDNA fragment.

One alternative method for cloning cDNA involves using synthetic DNA linkers which contain restriction sites. The double-stranded cDNA can be treated with *E. coli* DNA polymerase I or T4 DNA polymerase described above. The ds cDNA is made blunt ended by the $3' \rightarrow 5'$ exonuclease activity which removes the protruding 3' ends and the polymerase activity which fills in the recessed 3' ends. Linkers containing restriction enzyme sites can be blunt-end ligated to the ds cDNA by T4 DNA ligase. The ds cDNA can then be cloned into a vector using an appropriate restriction enzyme present in the DNA linker and on the vector. Use of different linkers at the two ends of the ds cDNA minimizes the danger of ds

Table 5
SPECIALIZED MICROBIAL VECTORS

Vector	Size (kb)[a]	Gene Markers	Single Restriction Site	Comments	Investigators	Ref.
Charon λ4A	45.4	Aam32, BamI, lac5, bio256, ∇KH54, ∇NIW5, φ80 QSR		Has been used for construction of libraries of 15—20 kb eukaryotic DNA fragments. On Xgal plates, blue plaques are formed.	Blattner (1977)	26
Charon λL47.1	40.6	(SrλI 1—2)∇, imm 434cI, NIN5, ChiA131		Can clone large EcoRI, HindIII and Bam4 fragments.	Loenen and Brammer (1980)	27
M13 mp8	7.2	β-gal2	BamHI, EcoRI, HindIII, PstI, SalI, SmaI	Polylinker contains the six enzyme sites for cloning. Insertion at any one of the sites results in inactivation of β-galactosidase.	Messing and Vieira (1982)	28
Cosmid pHc79	6.4	Ap^R, Tc^R	BamHI(Tc^R), ClaI, EcoRI, HindIII(Tc^R), PstI(Ap^R), SalI(Tc^R)	Can clone large DNA fragments (35—40 kb).	Hohn and Hinnen (1980)	29
Yeast Cosmid pYcl	10.0	Ap^R, his3 (yeast)	BamHI, EcoRI, SalI, XhoI	Can replicate in E. coli and yeast. Can clone large DNA fragments. (35—40 kb).	Hohn and Hinnen (1980)	29

[a] Kilobases.

cDNA self-circularization (Kurtz and Nicodemus, 1981).[36] To generate the largest population of ds cDNA containing two different linkers, one linker can be added before S1 nuclease treatment. Since the ds cDNA may contain sites for the restriction enzyme which would cut in the DNA linker sequence, the ds cDNA can be treated with a DNA modifying enzyme for the appropriate restriction enzyme. For example, if EcoRI was to be used, the cDNA could be treated with EcoRI methylase before blunt-end ligation with the DNA linker containing an EcoRI site (Maniatis et al., 1982).[4]

Another approach to preventing ds cDNA self-ligation when two similar synthetic linkers are used is to treat the ds cDNA with alkaline phosphatase which removes the 5' phosphates from both ends of the linear ds cDNA. The ds cDNA can ligate to the restricted vector which would have 5'-terminal P resulting in an open circular molecule with two nicks which will be repaired in the bacteria after transformation (Seeburg et al., 1977 and Ullrich et al., 1977).[32,37]

The most recent improvement for cDNA synthesis and cloning involves the use of specialized plasmids containing several strategically located restriction endonuclease sites which permit the addition of poly-T's to the plasmid which serves as a template for the poly-A containing mRNA (Okayama and Berg, 1982).[38] This procedure permits the synthesis of the cDNA as part of the vector and avoids the use of the S1 nuclease which is a major nemesis for efficient cDNA synthesis and cloning. As indicated in Figure 3, the Okayama and Berg system involves the following basic steps. Initially a plasmid primer and an oligo-dG-tailed linker DNA is prepared. The plasmid primer is digested with KpnI and oligo-dT tails are added. Treatment with HpaI removes one of the oligo-dT tails leaving the plasmid intact with only one oligo-dT tail which is used for mRNA annealing. cDNA is synthesized by terminal transferase using the T-tail of the plasmid which is annealed to the poly-A of the mRNA. Upon completion of the cDNA, oligo-dC tails are added to the 3' end of cDNA and the vector. The C-tail of the vector is removed by HindIII treatment and the C-tailed cDNA plasmid complex is ligated with a oligo(dG)-tailed linker which is prepared earlier from a different plasmid. The oligo-G linker also has a HindIII end which ligates with the HindIII end of the vector. The mRNA is degraded by RNase H and the second DNA strand is made by *E. coli* DNA polymerase I. The DNA ligase then ligates the cDNA creating a closed vector containing the cDNA.

The Okayama and Berg system results in a high percentage of full-length cDNA negating the need for cDNA fractionation on gradients if full-length cDNA is desired. In addition, the efficiency of cDNA synthesis and cloning is 1 to 3×10^5 transformants per µg starting mRNA (Okayama and Berg, 1982).[38] This efficiency is about 30 times that of standard procedures.

B. Gene Libraries

1. cDNA Libraries

Utilizing the various strategies for cDNA synthesis and cloning, it is possible to make a cDNA library or population of bacterial clones containing different cDNA made from a mixture of mRNA. Since most mammalian cells contain between 10,000 to 30,000 different mRNA species (Davidson, 1976),[39] it is possible to generate a population of bacterial clones which should contain cDNA representative of the whole mRNA population. To determine the number of cDNA clones representative of the whole mRNA population, the following formula can be utilized (Williams, 1981).[40]

$$N = \frac{\ln(1-P)}{\ln(1-1/n)}$$

N = number of clones required
P = probability desired (i.e., 0.99)
n = fraction of total mRNA population
  that specific mRNA represents (i.e., 1/30,000)

Enrichment of specific DNA clones can be obtained by size fractionation of poly A-RNA in sucrose density gradient centrifugation or glyoxal denaturing gel electrophoresis (McMaster and Carmichael, 1977).[41] Identification of desired mRNA fractions can be determined by probe hybridization, which will be discussed later, or by translation in frog oocytes (Gurdon et al., 1971),[42] or by in vitro translation systems. Desired Poly A-RNA fractions can then be pooled and used for cDNA synthesis. Isolation of clones containing cDNA made against very low abundance mRNAs depends on the generation of a sufficient number of clones according to the criteria described above and a strong screening procedure which will be described later.

2. Genomic Library

Genomic libraries contain cloned DNA isolated from cell genomes in contrast to synthesizing cDNA from mRNA. A major advantage of genomic libraries is that given enough clones one can generate a library with a 99% chance of containing any gene (Maniatis et al., 1982).[4]

$$N = \frac{\ln(1-P)}{\ln(1-f)}$$

N = number of clones required
P = desired probability (99%)
f = fraction of the genome represented
  in a single recombinant (must know
  size of genome and average size of
  the cloned inserts)

To generate a representative population of DNA-containing clones, the genomic DNA is usually mechanically sheared to average sizes of 16 to 20 kb and cloned into one of the appropriate Charon vectors using linkers (Maniatis et al., 1978).[43] Alternatively, partially digested genomic DNA can be cloned using specific restriction endonucleases. DNA fragments of 16 to 20 kb can be enriched by sucrose density gradients.

Uses of genomic libraries include walking along chromosomes since the library contains overlapping sequences, isolation of large DNA probes for cDNA cloning, screening, and isolation of genes for analysis or expression. Genomic sequences can be used in expression systems for production of human proteins if the gene doesn't contain an intron, isn't a pseudogene containing translational termination codons, and if some means are available for identifying the structural gene regions of the cloned DNA. One of the major tasks in isolating genes from genomic libraries involves screening, which we have optimized for cloning low-copy gene sequences.

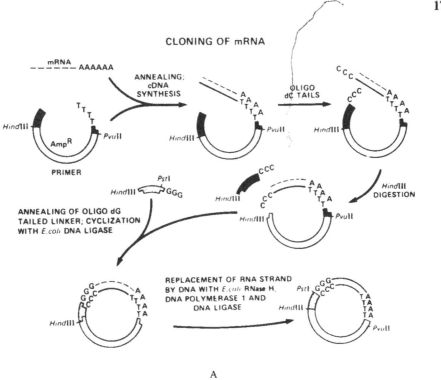

CLONING OF mRNA

A

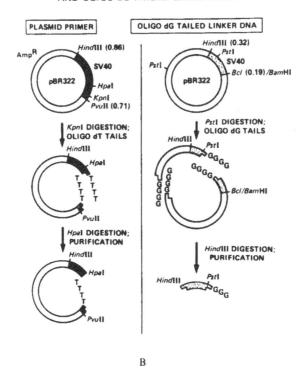

PREPARATION OF PLASMID PRIMER
AND OLIGO dG-TAILED LINKER DNA

B

FIGURE 3.   Cloning of mRNA using plasmid primer and oligo-dG-tailed linker DNA. (From Okayama, H., and Berg, P., *Mol. Cell. Biol.,* 2, 161, 1982. With permission.)

V. SCREENING AND ENRICHMENT OF DNA CLONES

A. Enrichment of DNA Clones

1. Chimeric Plasmid Enrichment

Clones of bacteria containing inserted sequences can be identified by insertional inactivation. For example, insertion of foreign DNA into the Pst I site of plasmid pBR322 inactivates the ampicillin resistance gene. Clones containing intact pBR322 would be Tet[R], Amp[R] whereas clones containing pBR322 with inserts at the Pst site would be Tet[R], Amp[S]. Differential phenotypes also permit positive selection of clones containing inserted DNA. Inactivation of the Tet[R] gene located on pBR322 results in the Tet[S] which can be selected for by growth in media containing fusaric acid or quinaldic acid (Bochner et al., 1980).[44] In rare cases, insertion does not result in inactivation of antibiotic genes. One insertion into the Pst I site of the Amp[R] gene of pBR322 resulted in a fused protein containing beta-lactamase activity (Villa-Komaroff et al., 1978).[34] An alternative method for enrichment of clones containing chimeric plasmid involves treatment of plasmid with alkaline phosphatase before ligation with DNA to be cloned, as described above (Seeburg et al., 1977 and Ullrich et al., 1977).[33,37]

2. DNA Size Enrichment

As previously described, sheared, or restricted DNA can be sized on sucrose density gradients before they are cloned. In addition to size separation on density gradients, DNA species can be separated by RPC-5 reverse-phase chromatography (Edgell et al., 1979).[45] The RPC-5 enrichment system is based on retention by the resin of DNA fragments such as restriction fragments due to A-T content, size, presence of single-strand ends, and other factors. Five to twentyfold enrichments of specific fragments are obtained as determined by assaying specific fractions with labeled probes such as beta-globin sequences. An extension of the RPC-5 procedure is to subject the DNA fractions containing desired sequences to preparative gel electrophoresis where a combined enrichment of 320-fold was achieved for beta-globin (Edgell et al., 1979).[45]

3. DNA Sequence Enrichment

Enrichment of specific DNA sequences can be achieved in cases where the restriction enzyme map for the desired gene is already known. This approach would involve the combination of selective restriction enzyme cleavage of the starting DNA material coupled with preparative electrophoresis of predicted restriction fragments which are then cloned with appropriate vectors. Since the purity of DNA fragments isolated from gels is important for subsequent cloning, low-melting agarose containing hydroxyethyl groups has been utilized to generate good yields of purified DNA fragments (Weislander, 1979).[46]

In the case of cDNA sequence enrichment, poly-A RNA can be fractionated by either sucrose density gradients or RNA gel electrophoresis. The gradient fractions or the gel slices can be assayed for specific mRNA by hybridization to appropriate probes for the desired gene, or by oocyte or in vitro translation as previously described. The RNA fractions containing the desired RNA species are then used for cDNA synthesis.

RNA enrichment can be achieved by immunoabsorption of specific polysomes containing the desired mRNA for cDNA syntheis by immunoaffinity columns (Schutz et al., 1977),[47] or protein-A Sepharose® columns (Shapiro and Young, 1981).[48] The combination of protein-A Sepharose and monoclonal antibodies has permitted 2 to 3 $\times$ 10[3]-fold enrichment of mRNA species (Korman et al., 1982).[49]

B. Screening

The identification and isolation of the clones containing specific genes ultimately involves screening genomic or cDNA libraries, even though gene or RNA enrichment procedures are employed. One of the most powerful techniques is to screen bacterial clones (Grunstein and Hogness, 1975),[50] or phage plaques (Benton and Davis, 1977),[51] with labeled probes consisting of either natural DNA fragments or synthetic DNA fragments based on known protein sequence information. In addition, RNA can be used for screening, utilizing differential hybridization strategies.

1. Synthetic Probes

a. Protein-Probe Strategy

Analysis of proteins by two-dimension electrophoresis (Anderson et al., 1978)[52] permits identification of rare proteins which are modulated by various biological processes, such as induction by viruses, mitogens, or hormone action. Even in cases where labeling of proteins is not convenient we have been able to detect 1 ng of protein by silver staining (Sammons et al., 1981).[53]

Following the identification of interesting proteins, their analysis can be extended by automated microsequencing which requires minute amounts of protein (5 pmol) which may exist as separated protein spots on two-dimensional gel electrophoresis (Hunkapillar and Hood, 1983).[54] The partial or complete amino acid sequence of the proteins can then be utilized as the basis for the synthesis of short DNA probes, 17 to 20 bases, which can be utilized for screening clones.

In choosing the region of the protein to serve as the basis for the probe sequence several factors are involved. Probes which are 17 to 20 bases and rich in G-C bases can be used for screening cDNA or genomic libraries. We have found that at least two sets of probes made against different parts of a gene are best. Due to degeneracy of the genetic code, protein regions rich in amino acids specified by a single codon (methionine-AUG; tryptophane-UGG) or two codons (histidine-CAU, CAS; phenylalamine-UUU, UUC; tyrosine-UAU, UAC) are preferred. In addition, one can choose codons that are preferentially utilized for specific amino acids.

The probes can be synthesized by the phosphodiester method (Agarwal et al., 1972)[55] or the more efficient and more popular phosphotriester method (Itakura et al., 1975).[56] The phosphodiester method required about 3×10^3 minutes of labor per internucleotide band formed, whereas the phosphotriester method only required 10^2 min/band, (Kaplan, personal communication).[5] Coupling the phosphotriester method with solid-phase synthesis has resulted in efficient techniques for synthesis of DNA probes. In fact, whole genes such as alpha interferon have been synthesized (Edge et al., 1981).[57]

b. Hybridization Conditions

Synthetic DNA has been utilized as primer in reverse transcription reactions with poly-A RNA for the synthesis of larger single-stranded probes for subsequent screening of cDNA libraries (Chan et al., 1979 and Goeddel et al., 1980a).[58,59] In addition, synthetic DNA probes of 14 to 20 bases have been used for successful screening of cDNA libraries (Montgomery et al., 1978, Goeddel et al., 1980a, and Wallace et al., 1981).[60,62] In our laboratory we have successfully screened a human genomic library with two 17-base probes containing sequences for alpha interferon and identified three clones containing alpha interferon related sequences. Fundamental to the effective use of synthetic probes are the proper hybridization conditions which optimize the positive hybridization signal against nonspecific background.

Optimization of hybridization conditions has been addressed by Wallace et al., (1981),[62] using two 14-base long probes for rabbit beta-globin DNA. One probe was exactly complementary to the beta-globin DNA and another 14-base probe had a one-base mismatch

(T→C). Using the filter hybridization procedure, conditions were established for hybridization with only the perfect-match probe. In addition, they used a mixture of 13-base long oligonucleotides representing eight of the possible coding sequences for the amino acids 15 to 19 encoded by the beta-globin DNA. Although only one of the eight probes was completely complementary, the mixture did hybridize to the beta-globin/DNA even under conditions where a one-base mismatch would not. Hence dilution of the exact copy probe by the seven one-base mismatch probes did not prevent detection of the hybridization signal (Wallace et al., 1981).[62] The stringency of the hybridization conditions was established by determining the Td, which is defined as the temperature at which half of the probe is dissociated (Wallace et al., 1979).[63]

We have optimized hybridization conditions using probes which are homologous to portions of the rat 18S ribosomal RNA gene (Torczynski, et al., 1983).[64] A mixture of probes primarily consisting of the 11-mer, 5′CGCAGTTTCAC3′, and the 13-mer, 5′TTCGCAGTTTCAC3′, were synthesized in our laboratory with a Bachem Gene Machine and were used in Northern, Southern, and dot-blot hybridizations against total human Namalwa cell RNA and plasmid DNA containing cloned rat 18S rRNA gene.

As can be seen in Figure 4, the sensitivity of the DNA dot blots was determined by hybridizing P^{32}-labeled probes described above with a 2.34 kb fragment of plasmid pDF-15 which contains part of the 18S rRNA gene including sequences homologous to the probes. Utilizing the hybridization conditions described in Figure 4 legend, 500 pg of DNA was detectable in a 16 hr exposure and 50 pg were detectable after 72 hr of exposure. RNA dot-blot hybridization was done with 18S rRNA isolated by sucrose density centrifugation and the same P^{32}-labeled probe. As can be seen in Figure 5, utilizing the hybridization conditions described, about 100 pg of 18S RNA was detectable after a 2-day exposure with enhancer screen at −80°C. The specificity of the probes is indicated in Figure 6 by lack of hybridization with 28S rRNA up to 25 ng and retention of strong hybridization between the probes and 18S rRNA even in the presence of excess 28S rRNA and tRNA. Concerning Northern hybridization of labeled probes with total RNA, the addition of dextran sulfate resulted in a major reduction in extraneous background resulting in clear 18S rRNA hybridization signals as indicated in Figure 7.

c. Screening a Human Genomic Library for Alpha-Interferon Genes Using Two Synthetic Probes

We have optimized the use of synthetic probes for the screening of human genomic libraries. Two probes for alpha-interferon were synthesized; one by Biologicals, Inc., and one in our laboratory using a Bachem Gene Machine. The 17-base sequences of each probe were based on sequence information of eight cDNA interferon clones (Goeddel et al., 1981).[68] For each of the probes complementary sequences were present in at least four of the eight described cDNA alpha-interferon genes whereas for three of the other genes there was only one base mismatch and two base mismatches for the last gene. Alpha-interferon sequences homologous to the two probes were about 50 bases apart. The sequences of the probes were

A   B

5′CAGCCAGGATGGAGTCC3′   5′CCTCCCAGGCACAAGGG3′

About 10,000 phage/80 cm^2 plate were screened separately with each probe as described in the legend to Figure 8. The phages were hybridized with each probe separately and positive clones were recognized as those clones which were positive for both probes.

From about 180,000 phages screened, five appeared to hybridize with both probes although several hybridized to either one of the two probes or were artifacts. Upon further analysis of the five phage clones, it was determined that three hybridized to both probes, one

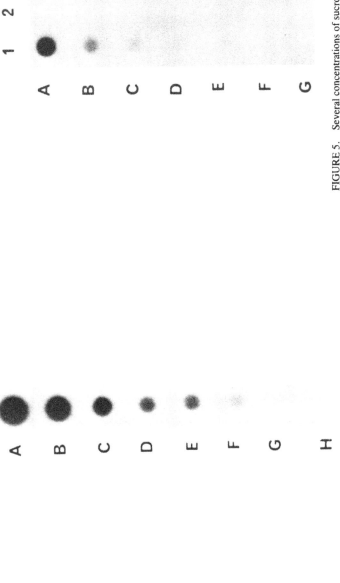

FIGURE 5. Several concentrations of sucrose gradient purified 18S (lane 1) and 28S (lane 2) rRNA. A = 25 ng; B = 5 ng; C = 2 ng; D = 1 ng; E = 0.5 ng; F = 0.1 ng; G = 0.05 ng; was hybridized at 37°C with 7 × 10⁵ dpm/mℓ 18S rRNA gene probe. The 18S and 28S rRNA was denatured according to White, B. A., and Bancroft, F. C., 1982,[67] blotted onto BA-85 filter with a BRL Hydri-dot apparatus, and baked at 70°C for 2 hr. Prehybridization was for 1 hr at 37°C in 6× NET, 5× Denhardt's, 0.5% SDS and 10% dextran SO₄. The filter was washed three times for 15' in a total volume of 150 mℓ containing 6× SSC and 0.1% SDS. The X-ray film was exposed to the filter for 2 days with enhancer screen at −80°C.

FIGURE 4. A 2.43 kb DNA fragment of the pDF−15 rat rRNA gene clone, Fuke et al., 1981,[65] was spotted onto nitrocellulose filter paper according to Thomas, P., 1980,[66] in the following concentrations: A = 10 ng; B = 5 ng; C = 2 ng; D = 1 ng; E = 0.5 ng; F = 0.2 ng; G = 0.1 ng; H = 0.05 ng. The filter was prehybridized for one hour at 37°C in 6× NET, 5× Denhardt's, 0.5% SDS, 10% dextran sulfate. The filter was then hybridized overnight with 8.1 × 10⁵ dpm/mℓ 18S rRNA gene probe at 32°C. The filter was washed three times at 37°C in a total volume of 150 mℓ containing 6× SSC and 0.1% SDS.

FIGURE 7. 1% agarose gel, glyoxalated RNA transferred to nitrocellulose filter according to Thomas, P., 1980.[66] Prehybridization, hybridization, and washing conditions were the same as in Figure 6 except 9 × 10[5] dpm/mℓ 18S rRNA gene probe was used. Lane a is without 10% dextran sulfate and Lane b is with 10% dextran sulfate. The location of the 28S rRNA and the 18S rRNA is indicated by 28 and 18, respectively.

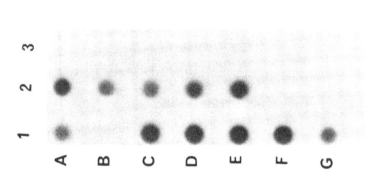

FIGURE 6. RNA dot hybridizations utilizing identical conditions as in Figure 5 except: Lane 1: A, 15 ng 18S rRNA alone; B, 15 ng 28S rRNA alone; C to G, 15 ng 18S rRNA + 15, 30, 75, 150, 300 ng 28S rRNA, respectively; Lane 2: A–E, 15 ng 18S rRNA + 150, 300, 750, 1500, 4500 ng rRNA, respectively; Lane 3: A to E, 150–4500 ng tRNA alone.

A   B

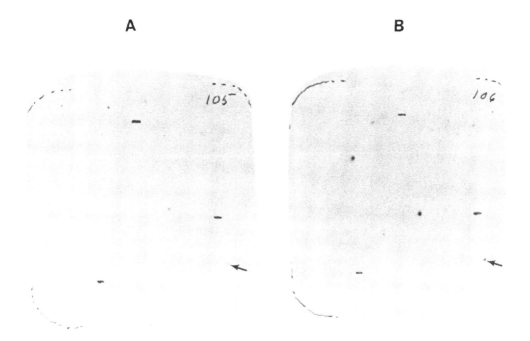

FIGURE 8. The arrows indicate positive phage screened with alpha-interferon probe 1 (A) and probe 2 (B), The three lines shown in the figures were used for orientation. The procedure involved screening of charon 4A human gene library, Lawn et al., 1978,[69] using *E. coli* host DP50 sup f. Overnight culture of DP50 was infected with phage at 37°C for 20′. Cells were plated at a density of 10,000 phage/80 cm^2 plate and grown for 5.5 to 6 hr. The plates were chilled and two copies per plate were made using nitrocellulose filters (Schleicher & Schuell BA85 SD). Blotted filters were denatured on 3M filter paper saturated with 0.5 N NaOH and 1.5 *M* NaCl, neutralized on paper saturated with 1 *M* Tris 7.5, again neutralized on paper saturated with 0.5 M Tris 7.5 + 1.5 *M* NaCl. The filters were baked in vacuo at 80°C for 2 hr. Filters were prehybridized overnight as described in Figure 5, except 0.1% SDS was used instead of 0.5% SDS. The two 17-mer alpha-interferon probes were end labeled with gamma ATP-P^{32} to a specific activity of $3\text{-}5 \times 10^8$ cpm/μg, using polynucleotide kinase. The labeled probes were purified by G-25 gel filtration and filtered through 0.2μ Gelman acrodisc filters. The labeled probes were boiled for three minutes and added for hybridization at a final concentration of 2 ng/mℓ. Each of the probes were individually hybridized to one of the two copies of each plate screened. The filters were hybridized at 37°C for twenty hours and were washed four times with $6 \times$ SSC at room temperature, ten minutes per wash, and once with $6 \times$ SSC + 0.1% SDS at 37°C for 1 hr. The filters were air-dried, exposed for 64 hours at -70°C with two intensifying screens and Kodak XAR film.

hybridized to only one of the probes, and the fifth was an artifact. Critical for the success of this procedure was to maximize the signal to background ratio as indicated in Figure 8. Several factors were necessary for enhanced signal and reduced background: (1) use of two different probes, (2) filtration of the labeled probes before addition to the hybridization mixture, (3) use of a device to rotate filters during prehybridization and hybridization, and (4) use of 10% dextran sulfate in prehybridization and hybridization solutions.

Analysis of the three putative alpha-interferon clones reinforced the designated identity. As indicated in Table 6, restriction analysis of the three clones, λ77, λ85 and λ105 with restriction enzymes BamHI, DdeI and EcoRI, followed by Southern hybridization with alpha-interferon probes A and B indicate that the same fragments hybridized to both probes. DNA sequence analysis of two clones by the (Sanger et al., 1977)[6] dideoxy sequencing system utilizing the interferon probes as primers for the DNA sequence reactions indicates the homology between the new clones and with previously described alpha-interferons (Pestka et al., 1983).[70] The homology of two of the clones is indicated in Figure 9 which shows part of a DNA sequencing gel containing both clones. DNA sequencing data indicates that one of the three alpha-interferon clones is related to alpha-interferon L, while the other

Table 6
CHARACTERIZATION OF
CLONED λ-INTERFERON GENES[a]

		BamHI	Dde	EcoRI
Probe A	λ 77	>20,000	1080	1820
	λ 85	>20,000	1020	2070
	λ105	na	na	7000
Probe B	λ 77	>20,000	1080	1820
	λ 85	>20,000	1020	2070
	λ105	na	na	7000

[a]   The DNA's of the three clones, λ77, λ85, and λ105 were cleaved with the indicated restriction enzymes and hybridized (Southern, 1975),[114] with probes A and B. Represented are the sizes (base pairs) of the restriction fragments which hybridize with the P32-labeled probes.

alpha-interferon gene appears to be different from alpha-interferon genes previously described. The third cloned gene is also being analyzed by DNA sequence analysis.

In conclusion, these results testify to the power of the synthetic probe screening approach which has enabled us to clone several alpha-interferon genes from a human genomic library using two 17-mer probes. Further utility of genomic libraries employing synthetic probes should be fruitful.

2. Differential Hybridization

If a probe is not available for screening, clones can be correlated to biological phenomena such as induction of specific mRNA species by differential hybridization. One example of this application is to induce cells for a specific protein and isolate poly-A RNA from the induced and uninduced cells. The two poly-A RNA populations are then used for cDNA synthesis. Since most of the cDNA species would be shared by both populations, except for the cDNA made against the induced mRNA, they can be used to screen a cDNA library made against induced poly-A RNA. Utilizing this procedure the galactase-inducible genes of yeast have been cloned (St. John and Davis, 1979).[71] Differential hybridization has also been utilized for cloning developmentally regulated mRNA. The synthesis of cDNA to specific developmental stages generates a population of developmental stage probes which can be used for screening cDNA or genomic libraries (Timberlake, 1980).[72]

Alternatively, induced poly-A RNA assayed in frog oocytes for specific protein can be used for cDNA synthesis and induced poly-A RNA can be used to screen the induced cDNA library. Colonies which appear to hybridize to induced RNA can then be plated and their DNA transferred to filters for hybridization with labeled RNA. The filter is washed, leaving behind RNA which is preferentially retained by the cDNA. The retained RNA can then be isolated and injected into frog oocytes for synthesis of the induced protein. This procedure was utilized for the initial cloning of one of the alpha-interferon genes (Pestka, 1983).[70] Recently, the human interleukin-2 (Il-2) gene was cloned starting with induced poly-A RNA from human leukemic T-cell lines stimulated by concanavalin A (Taniguchi et al., 1983).[73] The induced poly-A RNA was fractionated by centrifugation and fractions containing Il-2 activity were identified by frog oocyte assays. The active poly-A RNA was used for cDNA synthesis and cloned. Colonies containing Il-2 cDNA were identified by hybridization of cDNA from various colonies with induced poly-A RNA which was subsequently recovered and assayed in frog oocytes for Il-2 activity. The frequency of the Il-2 specific colonies in the cDNA library was 1:2000 (Taniguchi et al., 1983).[73]

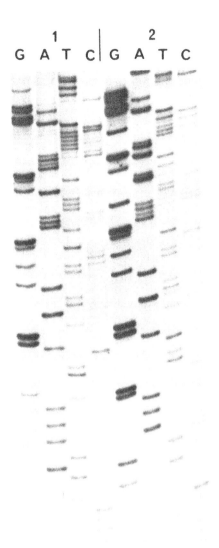

FIGURE 9. Sequencing gel of alpha-interferon gene demonstrating intergene homology. EcoRI fragments from lambda-phage clones were inserted into mp8 and DNA was sequenced using the di-deoxy method of Sanger et al., 1977.[6] The alpha-interferon probe A was used as a primer for the reaction.

3. Biological Activity and Immunological Screening

Although most of the screening techniques focus around the initial gene product, namely the RNA, systems have been developed which permit screening clones for protein products. Gene complementation has been employed in the cloning of bacterial, yeast, and *Neurospora* genes (Clarke and Carbon, 1976).[74] For shotgun cloning of yeast DNA into *E. coli* about 20% of the bacterial genes tested have been complemented by yeast genes. Obviously several factors are involved in such cross-species selection, promoter effectiveness, protein homology, and integrity of the cloned gene. Of course, transformation of yeast auxotrophs with yeast DNA has resulted in the isolation of several yeast genes (Strathern et al., 1982).[75] A variation of gene complementation has been the transformation of animal cells with DNA ligated to vectors and subsequent selection for altered cell function such as cell transformation. This technique has been employed by Weinberg and others to isolate human oncogenes (Weinberg, 1981 and Taparowsky et al., 1982).[77,78]

In addition to screening for biological activity, techniques have been established which permit detection of cross-reacting material (CRM) utilizing gene product-specific antibodies. This technique is useful in cases where cloned genes are expressed but the gene product is not active. One method employing radioimmunological screening involves the covalent attachment of antibody to CNBr-activated paper followed by the incubation of the paper with lysed colonies containing various chimeric DNA. The Ag-AB complex on the paper is then detected by incubation with ^{125}I-labeled antibody and autoradiography. The technique can detect 50 to 100 pg of antigen or gene product. Utilizing this technique (Clarke et al., 1979)[78] were able to detect from 6240 *E. coli* clones, 15 of which were positive for yeast hexokinase.

The coupling of biological activity or antibody screening with unique expression vectors permits an added dimension for screening for specific genes. A system has been developed by Guarente et al.,[79] which permits the cloning of foreign DNA between the lac promoter and 5′ end of the Z-gene which codes for beta-galactosidase resulting in fused protein retaining the enzyme activity. This technique utilizes the fact that a short N-terminal peptide of beta-galactosidase called the alpha-complementation fragment can be replaced with foreign protein and still retain beta-galactosidase activity. In this system, the foreign gene product furnishes the N-terminal end of the fused protein and therefore the AUG translational start codon. As can be seen in Figure 10, the foreign gene can be inserted in three reading frames permitting the selection for active beta-galactosidase. The lac promoter is added to the fused gene product and maximized expression of beta-galactosidase is achieved by a color reaction of the colonies for lactose fermentation (Guarente et al., 1980).[79] Following the identification of a clone with high lactose fermenting activity, the beta-galactosidase gene is removed and the rest of the foreign gene is inserted. At this point one can assay for the gene product by biological or immunological assays. Since the fused protein was highly active, lack of expression of the intact foreign gene should not reside in ineffective promotion or improper reading frame. Utilizing this technique, about 10,000 to 15,000 monomers per cell of rabbit beta-globin was achieved (Guarente et al., 1980).[79] This technique has been extended for analysis of genes cloned and expressed in yeast (Guarente, 1983).[80]

Expression selection has been extended to mammalian cells by (Okayama and Berg, 1983.)[81] They have developed a specialized expression vector which permits cDNA synthesis using oligo-dT primers attached to the vector as previously described (refer to Figure 3), but also contains a functional SV-40 promoter which can express foreign genes in animal cells. The SV-40 DNA component of the pcD vector ensures transcription, splicing and polyadenylation. Other promoters can be substituted for the SV-40 early region promoter. Clones isolated from human cDNA libraries made with the pcD vector were active in expressing hypoxanthine-guanine-phosphoribosyltransferase (Okayama and Berg, 1983).[81]

It is therefore possible to produce cDNA from RNA of cells which are induced for some activity or cells containing some developmental property and screen cDNA clones either by differential hybridization and/or expression of proteins by a biological or immunological assay. The ideal vector for such cloning would depend on which host cell would produce a product most similar to the native protein.

VI. EXPRESSION

The expression of foreign genes in host organisms such as bacteria (Table 7), yeast, or mammalian cells requires vectors which contain specific control sequences governing transcription and efficient splicing or RNA polyadenylation, in the case of animal host systems.

Transcription in *Escherichia coli* requires at least two conserved regions based on comparisons of several bacterial promoters. One is located about -35 bp from the start site (Rosenberg and Court, 1979)[93] and the other, the Pribnow® box, is located about -10 bp.

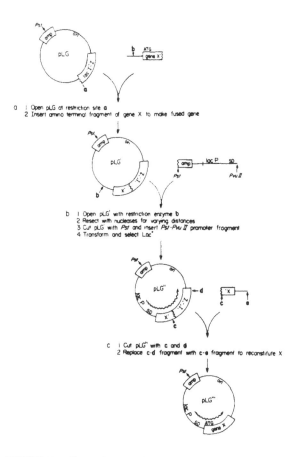

FIGURE 10. Expression selection utilizing the pLG plasmid containing part of the lac-operon which includes the lac promoter and part of the Z gene which codes for beta-galactosidase. As indicated, the inserted gene X is initially fragmented into N- and C-terminal fragments and sequentially inserted into the plasmid. (From Guarente, L., Roberts, T. M., and Ptashne, M., *Cell*, 20, 543, 1980. With permission.)

Mutation analysis of these two regions indicates their importance in modulating transcription efficiency. In addition, the distance between the two conserved sequences is important for maximizing promoter function (de Chrombrugghe et al., 1971 and Stefano and Grolla, 1982).[94,95] These conserved regions are probably involved in optimizing polymerase binding at the promoter region, permitting initiation of transcription. Transcription termination also involves signal regions, such as GC-rich regions followed by AT-rich sequences at the termination site which are probably recognized by RNA polymerase and rho termination factor (Rosenberg and Court, 1979).[93]

In addition to transcriptional control sequences, proper translation initiation in bacteria, such as *E. coli*, requires a ribosome binding site (rbs) (Shine and Dalgarno, 1975)[96] located about 3 to 12 bases upstream from the AUG translation initiation start site. Post-translational modification in bacteria involves removal of the formyl groups on the N-terminal methionine and possible removal of the N-terminal methionine. However, specific glycosylation and phosphorylation of eukaryotic proteins requires eukaryotic host-vector systems.

Transcription in eukaryotic organisms involves different promoter signal regions which are recognized by more complex RNA polymerase systems. In eukaryotic systems, one conserved regulatory region is the TATA control site which appears to be involved in correct

Table 7
EXPRESSION OF HUMAN GENES IN BACTERIA

Protein	Size (daltons)	DNA	Promoter	Product	Expression Level	Investigators	Ref.
Human Insulin	A = 2000 B = 3000	Chemical Synthesis	Lac	B-gal Fusion	10^5 mol/cell	Goeddel et al., 1979a	82
Human Growth Hormone	24,000	Chemical Synthesis	Lac-trp	rbs Fusion	1.8×10^5 mol/cell	Goeddel et al., 1979b	83
Human α-Interferon	20,000	Chemical Synthesis	Lac	rbs Fusion	10^7 μ/mℓ	De Maeyer et al., 1982	84
Human α-Interferon	20,000	cDNA	M13-Lac	B-gal Fusion	10^8—10^9 μ/ℓ	Slocombe et al., 1982	85
Human α-Interferon A	20,000	cDNA	trp	trp LE Fusion	1 μg/ℓ	Goeddel et al., 1980a	59
Human α-Interferon A	20,000	cDNA	trp	rbs Fusion	600 μg/ℓ	Shepard et al., 1982	86
Human α-Interferon B	20,000	cDNA	trp	trp Leader rbs Fusion	8×10^7 μ/ℓ	Yelverton et al., 1981	87
Human α-Interferon	20,000	cDNA	B-Lactamase	B-Lactamase Fusion	2 molecules/cell	Nagata et al., 1980	88
Human α-Interferon	20,000	Genomic DNA	M13-Lac	B-gal Fusion	10^7—10^8 μ/ℓ	Bollon et al.	115
Human β-Interferon	23,000 (Pre-H β-Ifn)	cDNA	Lac	rbs Fusion	50 molecules/cell	Taniguchi et al., 1980	89
Human β-Interferon	20,000	cDNA	Lac	rbs Fusion	2.2×10^3 mol/cell	Goeddel et al., 1980b	61
Human β-Interferon	20,000	cDNA	trp-trp-trp	trp Leader rbs Fusion	20×10^3 mol/cell	Goeddel et al., 1980b	61
Human γ-Interferon	17,000	cDNA	trp	trp Leader rbs Fusion	80 molecules/cell	Gray et al., 1982	90
Leu-enkephalin	600	Chemical Synthesis	Lac	B-gal Fusion	2×10^5 molecules/cell	Shemyakin et al., 1980	91
α-neo-endorphin	1,200	Chemical Synthesis	Lac	B-gal Fusion	5×10^5 molecules/cell	Tanaka et al., 1982	92

transcriptional initiation at the promoter site, and enhancer sequences which enhance transcription initiation and can exist some distance from the promoter regions (Breathnach and Chambon, 1981).[97] As indicated above, expression in mammalian cells using the Okayama and Berg[81] expression system involves the SV40 promoter sequence as well as sequences which permit RNA splicing and polyadenylation. Although several promoter systems have been utilized for expressing foreign genes, some have advantages in terms of their potential for control and levels of expression achieved.

Components of the Lac operon have been utilized for the expression of several genes in bacteria as indicated in Table 7. The Lac promoter has been used either alone (Goeddel et al., 1979)[82] or in concert with a Trp promoter (Goddel et al., 1979)[83] employing vectors such as pBR322 or M13 (see Tables 3 and 5 and Slocombe et al., 1982[85]). The protein product synthesized via a lac promoter is generated either as a unique protein (DeMaeyer et al., 1982)[84] or as a fusion product with beta-galactosidase which can then be cleaved with CNBr (Goeddel et al., 1979).[82] If methionine is present in the foreign protein, other proteolytic techniques must be employed (Shine et al., 1980).[98] The Lac system has the advantage of being regulated by Lac induction or catabolite repression. In addition, the generation of beta-gal fusion products permits positive selection of high producers of foreign gene products due to the activity of beta-galactosidase which renders blue colonies on agar plates containing Xgal, as described in Figure 10 (Guarente et al., 1980).[79] In addition, fusion products can be protected from proteolytic attack which occurs with some foreign proteins or small peptides (Itakura et al., 1977).[99] High production of human alpha-2 interferon (10^8 to 10^9 μ/ℓ) has been achieved employing the M13-Lac system (Slocombe et al., 1982).[83]

The Trp promoter is an efficient control element which can result in the generation of high levels of trp enzymes when the operon is present on a plasmid vector. About 25% of total bacterial protein can be trp enzymes when the operon is induced with 3-indolylacrylic acid (Yanofsky et al., 1981).[100] Initial trp expression vectors contained the trp promoter, operator, leader and attenuator, the trp E gene, and part of the trp D gene inserted in to pBR322. Further modifications resulted in the production of the trp E gene product at a level comprising 30% of total cell protein. The trp D fragment was also made and was stable, perhaps due to the fact that a trp D-E hybrid protein was generated (Hallelwell and Emtage, 1980).[101]

As can be seen in Table 7, the trp promoter system has been used for the production of human alpha-interferon (Goedell et al., 1980 and Shepard et al., 1982)[59,86] and human beta-interferon either as a single promoter element or as a set of three tandem promoters. The three trp promoter system resulted in the production of 20×10^3 molecules/cell of human beta-interferon, which was ten-fold higher than comparable production with a single Lac promoter (Goeddel et al.,1980).[61]

The *lambda PL promoter* is a strong control element which controls early leftward transcription of lambda genes N-int (Szybalski and Szybalski, 1979).[102] The PL promoter is repressed by the CI product and by the CRO product. A gene coding for a temperature-sensitive CI product λCI ts B57 can be used in the vector or in the host for controlled expression of foreign genes (Bernard et al., 1979).[103] The PL promoter has been employed either in lambda or plasmid vectors for expression of foreign genes. The small t antigen of SV40 has been produced via a PL promoter preceded by a ribosome binding site. The t antigen was produced at a level comprising 42.5% of bacterial total protein which was higher than levels obtained utilizing a Lac promoter, although truncated t antigen was also produced (Roberts et al., 1979).[104]

The *beta-lactamase promoter* has been utilized for low-level expression of hybrid proteins such as rat pregrowth hormone made in bacterial minicells (Seeburg et al., 1978).[105] Beta-lactamase-insulin hybrid protein had been detected in cellular periplasmic spaces utilizing a solid-phase radioimmunoassay for selection of clones containing fused products (Broome

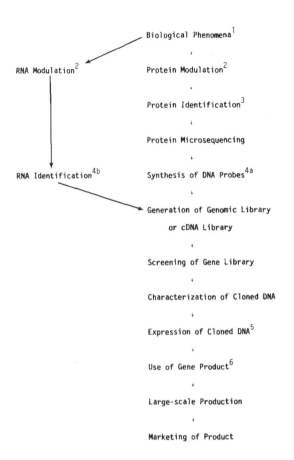

FIGURE 11.   (1) Such as antiviral action, anticancer action, morphogenesis, etc., (2) Such as induction of protein or RNA related to biological phenomena, (3) Differential analysis by 2D-gel electrophoresis, (4a) Based on protein sequence, (4b) Oocyte assay, (5) Correlate product activity to biological phenomena, (6) Analysis and preclinical-clinical trials.

and Gilbert, 1978).[106] Capitalizing on the beta-lactamase secretion capacity, several plasmids (pKT) have been developed and utilized for the production of human proinsulin which is also detectable in the periplasmic space (Talmadge et al., 1980).[107] A requirement for the secretion of the protein was the presence of a signal sequence which could be recognized by the bacterial signal peptidase.

In addition to developing more efficient promoter systems, increased expression of gene products can be achieved by increasing the dosage of the foreign gene. The gene and control elements such as the promoter, and ribosome-binding site can be transferred to plasmids that are temperature inducible, "runaway replication" vectors (pKN402).[108] A derivative of pKN402, the vector pA52, has been coupled with Lac or Trp promoters and foreign genes, resulting in a 1000-fold increase in vector and foreign gene copy number.[109]

VII. SUMMARY

This chapter contains a concise description of major techniques employed in the cloning and expression of genes thus permitting the large-scale production of useful proteins as summarized in Figure 11.

More detailed procedures can be obtained from the cited references, the laboratory manual on molecular cloning by Maniatis et al.,[4] and specialized monographs.[110-112]

Further treatment of the expression of genes can be found in Maniatis et al.,[4] and Harris,[113] and subsequent chapters of this book, including chapters on the use of yeast expression systems and preclinical and clinical trials of recombinant DNA products.

NOTES ADDED IN PROOF

The new human alpha interferon isolated from a genomic library as described in Section IV.B.1 has been expressed in *E. coli* using a M13-Lac expression vector. The level of production achieved was 10^7 to 10^8 μ/ℓ as indicated in Table 7.[115]

ACKNOWLEDGMENTS

Part of this work was performed in the Oree Meadows Perryman Molecular Genetics Laboratory for Cancer Research at Wadley Institutes with support from the Meadows Foundation and a research grant from the National Institutes of Health, GM28090, to APB. We thank Carol Crumley for her assistance in the generation of this chapter.

REFERENCES

1. **Roberts, T. M., Swanberg, S. L., Poteete, A., Riedel, G., and Backman, K.,** A plasmid cloning vehicle allowing a positive selection for inserted fragments, *Gene,* 12, 123, 1980.
2. **Kelly, R. G., Cozzarelli, N., Deutscher, M. P., Lehman, I. R., and Kornberg, A.,** Enzymatic synthesis of deoxyribonucleic acid, *J. Biol. Chem.,* 245, 39, 1970.
3. **Maniatis, T., Jeffrey, A., and Kleid, D. G.,** Nucleotide sequence of the rightward operator of phage lambda, *Proc. Natl. Acad. Sci.,* 72, 1184, 1975.
4. **Maniatis, T., Fritsch, E. F., and Sambrook, J.,** Molecular Cloning, a Laboratory Manual, *Cold Spring Harbor Laboratory Publications,* New York, 1982.
5. **Jacobsen, H., Klenow, H., and Ovargaard-Hansen, K.,** The N-terminal amino-acid sequences of DNA polymerase I from *Escherichia coli* and of the large and the small fragments obtained by a limited proteolysis, *Eur. J. Biochem.,* 45, 623, 1974.
6. **Sanger, F., Nicklen, S., and Coulson, A. R.,** DNA sequencing with chain-terminating inhibitors, *Proc. Natl. Acad. Sci.,* 74, 5463, 1977.
7. **Bollum, F. J.,** Terminal deoxynucleotidyl transferase, *Enzymes,* 10, 1974, 145.
8. **Tu, C.-P. D. and Cohen, S. N.,** 3-End labeling of DNA with [alpha-^{32}P] cordycepin-5'-triphosphate, *Gene,* 10, 177, 1980.
9. **Richardson, C. C.,** Polynucleotide kinase from *Escherichia coli* infected with bacteriophage T4, *Proc. Nucleic Acid Res.,* 2, 815, 1971.
10. **Maxam, A. M. and Gilbert, W.,** A new method for sequencing DNA, *Proc. Natl. Acad. Sci.,* 74, 560, 1977.
11. **Verma, I. M.,** The reverse transcriptase, *Biochim. Biophys. Acta,* 473, 1, 1977.
12. **Sugino, A., Goodman, H. M., Heyneker, H. L., Shine, J., Boyer, H. W., Cozzarelli, N. R.,** Interaction of bacteriophage T4 RNA and DNA ligases in joining of duplex DNA at base-paired ends, *J. Biol. Chem.,* 252, 3987, 1977.
13. **Little, J. W., Lehman, I. R., and Kaiser, A. D.,** An exonuclease induced by bacteriophage lambda, *J. Biol. Chem.,* 242, 672, 1967.
14. **Hershfield, V., Boyer, H. W., Yanofsky, C., Lovett, M. A., and Helsinki, D. R.,** Plasmid ColE1 as a molecular vehicle for cloning and amplification of DNA, *Proc. Natl. Acad. Sci.,* 71, 3455, 1974.
15. **Bolivar, F., Rodriguez, R. L., Greene, P. J., Betlach, M. C., Heynecker, H. L., Boyer, H. W., Crosa, J. H., and Falkow, S.,** Construction and characterization of new cloning vehicles. II. A multipurpose cloning system, *Gene,* 2, 95, 1977.

16. **Sutcliffe, J. G.**, pBR322 restriction map marked from the DNA sequence: accurate DNA size markers up to 4361 nucleotide pairs long, *Nucleic Acids Res.*, 5, 2721, 1978.

17. **Soberon, X., Covarrubias, L., and Bolivar, F.**, Construction and characterization of new cloning vehicles. III. Derivatives of plasmid pBR322 carrying unique EcoRI sites for selection of EcoRI generated recombinant DNA molecules, *Gene*, 4, 121, 1978.

18. **Rao, R. N. and Rogers, S. G.**, Plasmid pKC7: a vector containing ten restriction endonuclease site suitable for cloning DNA segments, *Gene*, 7, 79, 1979.

19. **Hinnen, A., Hicks, J. B., and Fink, G. R.**, Transformation of yeast, *Proc. Natl. Acad. Sci.*, 75(4), 1929, 1978.

20. **Bollon, A. P. and Silverstein, S. J.**, Techniques of DNA-mediated gene transfer for eucaryotic cells, in *Techniques in Somatic Cell Genetics*, Shay, J. W., Ed., Plenum Press, New York, 1982, 415.

21. **Botstein, D. and Davis, R. W.**, Principles and practice of recombinant DNA research with yeast, in *The Molecular Biology of the Yeast Saccharomyces cerevisiae — Metabolism and Gene Expression*, Strathern, J. N., Jones, E. W. and Broach, J. R., Eds., Cold Spring Harbor Laboratory Press, New York, 1982, 607.

22. **Broach, J. R., Strathern, J. N., and Hicks, J. B.**, Transformation in yeast: development of a hybrid cloning vector and isolation of the Can1 gene, *Gene*, 8, 121, 1979.

23. **Stinchcomb, D. T., Struhl, K., and Davis, R. W.**, Isolation and characterization of a yeast chromosome replicator, *Nature*, 282, 39, 1979.

24. **Barron, E. A., Sidhu, R. S., and Bollon, A. P.**, Construction of a yeast-bacterial DNA cloning vector, *J. Clin. Hem. Oncol.*, in press, 1983.

25. **Clarke, L. and Carbon, J.**, Isolation of the centromere-linked CDC10 gene by complementation in yeast, *Proc. Natl. Acad. Sci.*, 77(4), 2173, 1980.

26. **Blattner, F. R., Williams, B. G., Blechl, A. E., Denniston-Thompson, K., Faber, H. E., Furlong, L.-A., Grunwald, D. J., Kiefer, D. O., Moore, D. D., Sheldon, E. L., and Smithies, O.**, Charon phages: safer derivatives of bacteriophage lambda for DNA cloning, *Science*, 196, 161, 1977.

27. **Loenen, W. A. and Brammar, W. J.**, A bacteriophage lambda vector for cloning large DNA fragments made with several restriction enzymes, *Gene*, 10, 249, 1980.

28. **Messing, J. and Vieira, J.**, A new pair of M13 vectors for selecting either DNA strand of double-digest restriction fragments, *Gene*, 19, 269, 1982.

29. **Hohn, B. and Hinnen, A.**, Cloning with cosmids in *E. coli* and yeast, *Genetic Engineering*, Vol. 2, Setlow, J. K. and Hollender, Eds., Plenum Press, N.Y., 1980, 169.

30. **Collins, J. and Hohn, B.**, Cosmids: a type of plasmid gene-cloning vector that is packagable *in vitro* in bacteriophage h heads, *Proc. Natl. Acad. Sci.*, 75, 4242, 1978.

31. **Slocombe, P., Easton, A., Boseley, P., and Burke, D. C.**, High level expression of an interferon alpha-2 gene cloned in phage M13 mp7 and subsequent purification with a monoclonal antibody, *Proc. Natl. Acad. Sci.*, 79, 5455, 1982.

32. **Efstratiadis, A., Kafatos, F. C., Maxam, A. M., and Maniatis, T.**, Enzymatic *in vitro* synthesis of globin genes, *Cell*, 7, 279, 1976.

33. **Seeburg, P. H., Shine, J., Marshall, J. A., Baxter, J. D., and Goodman, H. M.**, Nucleotide sequence and amplification in bacteria of structural gene for rat growth hormone, *Nature*, 220, 486, 1977.

34. **Villa-Komaroff, L., Efstratiadis, A., Broome, S., Lomedico, P., Tizard, R., Naker, S. P., Chick, W. L., and Gilbert, W.**, A bacterial clone synthesizing proinsulin, *Proc. Natl. Acad. Sci.*, 75, 3727, 1978.

35. **Peacock, S. L., McIver, C. M. and Monohan, J. J.**, Transformation of *E. coli* using homopolymer-linked plasmid chimeras, *Biochim. Biophys. Acta*, 655, 243, 1981.

36. **Kurtz, D. T. and Nicodemus, C. F.**, Cloning of alpha-2-micron globulin DNA using a high efficiency technique for the cloning of trace messenger RNAs, *Gene*, 13, 145, 1981.

37. **Ullrich, A., Shine, J., Chirgwin, J., Pictet, R., Tischer, E., Rutter, W. J., and Goodman, H. M.**, Rat insulin genes: construction of plasmids containing the coding sequences, *Science*, 196, 1313, 1977.

38. **Okayama, H. and Berg, P.**, High-efficiency cloning of full-length cDNA, *Mol. Cell. Biol.*, 2, 161, 1982.

39. **Davidson, E.**, *Gene Activity in Early Development*, Academic Press, New York, 1976.

40. **Williams, J. G.**, The preparation and screening of a cDNA clone bank, in *Genetic Engineering*, Vol. 1, Williamson, R., Ed., Academic Press, New York, 1981, p.2.

41. **McMaster, G. K. and Carmichael, G. G.**, Analysis of single- and double-stranded nucleic acids on polyacrilamide and agarose gels by using glyoxal and acridine orange, *Proc. Natl. Acad. Sci.*, 74, 4835, 1977.

42. **Gurdon, J. B., Lane, C. D., Woodland, H. R., and Maraix, G.**, Use of frog eggs in oocytes for the study of mRNA and its translation in living cells, *Nature*, 233, 177, 1971.

43. **Maniatis, T., Hardison, R. C., Lacy, E., Lauer, J., O'Connell, C., Quon, D., Sim, D. K., and Efstratiadis, A.**, The isolation of structural genes from libraries of eucaryotic DNA, *Cell*, 15, 687, 1978.

44. **Bochner, B. R., Huang, H. C., Schieven, G. L., and Ames, B. N.**, Positive selection for loss of tetracycline resistance, *J. Bacteriol.*, 143, 926, 1980.

45. **Edgell, M. H., Weaver, S., Haigwood, N., and Hutchison, C. A., III** Gene enrichment, in *Genetic Engineering*, Vol. 1 Setlow, J. K. and Hollaender, A., Eds., Plenum Press, New York, 1979, p.37.

46. **Weislander, L.,** A simple method to recover intact high molecular weight RNA and DNA after electrophoretic separation in low gelling temperature agarose gels, *Anal. Biochem.,* 98, 305, 1979.

47. **Schutz, G., Kieval, S., Groner, B., Sippel, A. E., Kurtz, D. T., and Feigelson, P.,** Isolation of specific messenger RNA by adsorption to matrix-bound antibody, *Nucleic Acids Res.,* 4, 71, 1977.

48. **Shapiro, S. Z. and Young, J. R.,** An immunochemical method for mRNA purification, *J. Biol. Chem.,* 256, 1495, 1981.

49. **Korman, A. J., Knudsen, P. J., Kaufman, J. F., and Strominger, J. L.,** cDNA clones for the heavy chain of HLA-DR antigens obtained after immunopurification of polysomes by monoclonal antibody, *Proc. Natl. Acad. Sci.,* 79, 1844, 1982.

50. **Grunstein, M. and Hogness, D.,** Colony hybridization: a method for the isolation of cloned DNAs that contain a specific gene, *Proc. Natl. Acad. Sci.,* 72, 3961, 1975.

51. **Benton, W. D. and Davis, R. W.,** Screening lambda-gt recombinant clones by hybridization to single plaques in situ, *Science,* 196, 180, 1977.

52. **Anderson, N. G. and Anderson, N. L.,** Analytical techniques for cell fractions XXI. Two-dimensional analysis of serum and tissue proteins: multiple isoelectric focusing, *Anal. Biochem.,* 85, 331, 1978.

53. **Sammons, D. W., Adams, L. D., and Wishigawa, E. E.,** Ultrasensitive silver-based color staining of polypeptides in polyacrylamide gels, in *Electrophoresis,* 2, 135, 1981.

54. **Hunkapillar, M. W. and Hood, L. E.,** Protein sequence analysis: automated microsequencing, *Science,* 219, 650, 1983.

55. **Agarwal, K.-L., Yamakazi, A., Fashion, P. J., and Khorana, H. G.,** Chemical synthesis of polynucleotides, *Agnew. Chem. Int. Ed. Engl.,* 11, 451, 1972.

56. **Itakura, K., Katagiri, N., Narang, S. A., Bahl, C. P., Marians, K. J. and Su, R.,** Chemical synthesis and sequence studies of deoxyribooligonucleotides which constitute the duplex sequence of the lactose operator of *Escherichia coli, J. Biol. Chem.,* 250, 4592, 1975.

57. **Edge, M. D., Greene, A. R., Heathcliffe, G. R., Meacock, P. A., Schuch, W., Scanlon, D. B., Atkinson, T. C., Newton, C. R., and Markham, A. F.,** Total synthesis of a human leukocyte interferon gene, *Nature,* 292, 756, 1981.

58. **Chan, S. J., Noyes, B. E., Agarwal, K. L., and Steiner, D. F.,** Construction and selection of recombinant plasmids containing full-length complementary DNAs corresponding to rat insulins I and II, *Proc. Natl. Acad. Sci.,* 76, 5036, 1979.

59. **Goeddel, D. V., Yelverton, E., Ullrich, A., Heynecker, H. L., Miozzari, G., Holmes, W., Seeburg, P. H., Dull, T., May, L., Stebbing, N., Crea, R., Maeda, S., McCandliss, R., Sloma, A., Tabor, J. M., Gross, M., Familletti, P. C., and Pestka, S.,** Human leukocyte interferon produced by *E. coli* is biologically active, *Nature,* 287, 411, 1980a.

60. **Montgomery, D. L., Hall, B. D., Gillam, S., and Smith, M.,** Identification and isolation of the yeast cytochrome c gene, *Cell,* 14, 673, 1978.

61. **Goeddel, D. V., Shepard, H. M., Yelverton, E., Leung, D., and Crea, R.,** Synthesis of human fibroblast interferon by *E. coli, Nucleic Acids Res.,* 8, 4057, 1980b.

62. **Wallace, R. B., Johnson, N. J., Hirose, T., Miyake, M., Kawashima, E. H., and Itakura, K.,** The use of synthetic oligonucleotides as hybridization probes. II. Hybridization of oligonucleotides of mixed sequence to rabbit beta-globin DNA, *Nucleic Acids Res.,* 9, 879, 1981.

63. **Wallace, R. B., Schaffer, J., Murphy, R. F., Bonner, J., Hirose, T., and Itakura, K.,** Hybridization of synthetic oligodeoxyribonucleotides to phi × 174 DNA: the effect of single base pair mismatch, *Nucleic Acids Res.,* 6, 3543, 1979.

64. **Torcynski, R., Bollon, A. P., and Fuke, M.,** The complete nucleotide sequence of the rat 18S ribosomal RNA gene and comparison with the respective yeast and frog genes, *Nucleic Acids Res.,* in press, 1983.

65. **Fuke, M., Dennis, K. J., and Busch, H.,** Characterization of cloned rat ribosomal DNA fragments, *Mol. Gen. Genet.,* 182, 25, 1981.

66. **Thomas, P. S.,** Hybridization of denatured RNA and small DNA fragments transferred to nitrocellulose, *Proc. Natl. Acad. Sci.,* 77, 5201, 1980.

67. **White, B. A. and Bancroft, F. C.,** Cytoplasmic dot hybridization, *J. Biol. Chem.,* 257, 8569, 1982.

68. **Goeddel, D. V., Leung, D. W., Dull, T. J., Gross, M., Lawn, R. M., McCandliss, R., Seeburg, P. H., Ullrich, A., Yelverton, E., and Gray, P. W.,** The structure of eight distinct cloned human leukocyte interferon cDNAs, *Nature,* 290, 20, 1981.

69. **Lawn, R. M., Fritsch, E. F., Parker, R. C., Blake, G., and Maniatis, T.,** The isolation and characterization of a linked delta- and beta-globin gene from a cloned library of human DNA, *Cell,* 15, 1157, 1978.

70. **Pestka, S.,** The purification and manufacture of human interferons, *Sci. Am.,* 249, 37, 1983.

71. **St. John, T. P. and Davis, R. W.**, Isolation of galactose-inducible DNA sequences from *Saccharomyces cerevisiae* by differential plaque filter hybridization, *Cell*, 16, 443, 1979.

72. **Timberlake, W. E.**, Developmental gene regulation in *Aspergillus nidulans*, *Dev. Biol.*, 78, 497, 1980.

73. **Taniguchi, T., Matsui, H., Fujita, T., Takaska, C., Kashima, N., Yoshimoto, R., and Hamuro, J.**, Structure and expression of a cloned cDNA from human interleukin 2, *Nature*, 302, 305, 1983.

74. **Clarke, L. and Carbon, J.**, A colony bank containing synthetic ColEI hybrid plasmids representative of the entire *E. coli* genome, *Cell*, 9, 91, 1976.

75. **Strathern, J. N., Jones, E. W. and Broach, J. R.**, *The Molecular Biology of the Yeast Saccharomyces — Metabolism and Gene Expression*, Cold Spring Harbor Laboratory Press, New York, 1982, 607.

76. **Weinberg, R. A.**, Use of transfection to analyze genetic information and malignant transformations, *Biochim. Biophys. Acta*, 651, 25, 1981.

77. **Taparowsky, E., Suard, Y., Fassano, O., Shimizu, K., Goldfarb, M., and Wigler, M.**, Activation of the T24 bladder carcinoma transforming gene is linked to a single amino acid change, *Nature*, 300, 762, 1982.

78. **Clarke, L., Hitzeman, R., and Carbon, J.**, Selection of specific clones from colony banks by screening with radioactive antibody, *Methods Enzymol.*, 68, 436, 1979.

79. **Guarente, L., Roberts, T. M., and Ptashne, M.**, Improved methods for maximizing expression of a cloned gene: a bacterium that synthesizes rabbit beta-globin, *Cell*, 20, 543, 1980.

80. **Guarente, L.**, Yeast promoters and Lac 2 fusions designed to study expression of cloned genes in yeast, in *Methods in Enzymology*, Vol. 101, Wu, R., Ed., Academic Press, New York, 1983, 181.

81. **Okayama, H. and Berg, P.**, A cDNA cloning vector that permits expression of cDNA inserts in mammalian cells, *Mol. Cell. Biol.*, 3, 280, 1983.

82. **Goeddel, D. V., Kleid, D. G., Bolivar, F., Heynecker, H. C., Yansura, D. G., Crea, R., Hirose, T., Kraszeuski, A., Itakura, K., and Riggs, A. D.**, Expression in *E. coli* of chemically synthesized genes for human insulin, *Proc. Natl. Acad. Sci.*, 76, 106, 1979a.

83. **Goeddel, D. V., Heynecker, H. L., Hozumi, T., Arentzen, R., Itakura, K., Yansura, D. G., Ross, M. J., Miozzari, G., Crea, R., and Seeburg, P. H.**, Direct expression in *Escherichia coli* of a DNA sequence coding for human growth hormone, *Nature*, 281, 544, 1979b.

84. **De Maeyer, E., Skup, D., Prasad, K. S. N., De Maeyer-Guignard, J., Williams, B., Sharpe, G., Pioli, D., Hennam, J., Schuch, W., and Atherton, K. T.**, Expression of chemically synthesized human alpha-interferon gene, *Proc. Natl. Acad. Sci.*, 79, 4256, 1982.

85. **Slocombe, P., Easton, A., Boseley, P., and Burke, D. C.**, High level expression of an interferon alpha2 gene cloned in phage M13 mp7 and subsequent purification with a monoclonal antibody, *Proc. Natl. Acad. Sci.*, 79, 5455, 1982.

86. **Shepard, H. M., Yelverton, E., and Goeddel, D. V.**, Increased synthesis in *E. coli* of fibroblast and leukocyte interferons through alterations in ribosome-binding sites, *DNA*, 1, 125, 1982.

87. **Yelverton, E., Leung, D., Weck, P., Gray, P. W., and Goeddel, D. V.**, Bacterial synthesis of a novel human leukocyte interferon, *Nucleic Acids Res.*, 9, 731, 1981.

88. **Nagata, S., Taira, H., Hall, A., Johnsrud, L., Streuli, M., Ecsodi, J., Boll, W., Cantell, K., and Weissmann, C.**, Synthesis in *E. coli* of a polypeptide with human leukocyte interferon activity, *Nature*, 284, 316, 1980a.

89. **Taniguchi, T., Guarente, L., Roberts, T. M., Kimelman, D., Douhan, J., and Ptashne, M.**, Expression of the human fibroblast interferon gene in *E. coli*, *Proc. Natl. Acad. Sci.*, 77, 5230, 1980.

90. **Gray, P. W., Leung, D. W., Pennica, D., Yelverton, E., Wajarian, R., Simonsen, C. C., Derynck, R., Sherwood, P. J., Wallace, D. M., Berger, S. L., Levinson, A. D., and Goeddel, D. V.**, Expression of human immune interferon cDNA in *E. coli* and monkey cells, *Nature*, 295, 503, 1982.

91. **Shemyakin, M. F., Chestukin, A. V., Dolganov, G. M., Khodkova, E. M., Monastyrskaya, G. S., and Sverdlov, E. D.**, Leu-enkephalin purification from *E. coli* cells carrying the plasmid with fused synthetic leu-enkephalin gene, *Nucleic Acids Res.*, 8, 6163, 1980.

92. **Tanaka, S., Oshima, T., Ohsue, K., Ono, T., Oikawa, S., Takano, I., Noguchi, T., Kangawa, K., Minamino, N., and Matsuo, H.**, Expression in *Escherichia coli* of chemically synthesized gene for a novel opiate peptide alpha neo-endorphin, *Nucleic Acids Res.*, 10, 1741, 1982.

93. **Rosenberg, M. and Court, D.**, Regulatory sequences involved in the promotion and termination of RNA transcription, *Ann. Rev. Genetics*, 13, 319, 1979.

94. **de Crombrugghe, B., Chen, B., Gottesman, M., Pastan, I., Varmus, H. E., Emmer, M., and Perlman, R.**, Regulation of *lac* mRNA synthesis in a soluble cell-free system, *Nat. New Biol.*, 230, 37, 1971.

95. **Stephano, J. E. and Gralla, J. D.**, Spacer mutations in the *lacP*ˢ promoter, *Proc. Natl. Acad. Sci.*, 79, 1069, 1982.

96. **Shine, J. and Dalgarno, L.**, Determinant of cistron specificity in bacterial ribosomes, *Nature*, 254, 34, 1975.

97. **Breathnach, R. and Chambon, P.**, Organization and expression of eucaryotic slit genes coding for proteins, *Ann. Rev. Biochem.*, 50, 349, 1981.

98. **Shine, J., Fettes, I., Lan, N. C. Y., Roberts, J. L., and Baxter, J. D.,** Expression of a cloned beta-endorphin gene sequence by *Escherichia coli, Nature,* 285, 456, 1980.

99. **Itakura, K., Hirose, T., Crea, R., Riggs, A. D., Heynecker, H. L., Bolivar, F., and Boyer, H. W.,** Expression in *E. coli* of a chemically synthesized gene for the hormone somatostatin, *Science,* 198, 1056, 1977.

100. **Yanofsky, C.,** Attenuation in the control of expression of bacterial operons, *Nature,* 289, 751, 1981.

101. **Hallewell, R. A. and Emtage, S.,** Plasmid vectors containing the tryptophan operon promoter suitable for efficient regulated expression of foreign genes, *Gene,* 9, 24, 1980.

102. **Szybalski, E. H. and Szybalski, W.,** A comprehensive molecular map of bacteriophage lambda, *Gene,* 7, 217, 1979.

103. **Bernard, H. U., Remaut, E., Hershfield, M. V., Das, H. K., Helinski, D. R., Yanofsky, C., and Franklin, N.,** Construction of plasmid cloning vehicles that promote gene expression from the bacteriophage lambda P_L promoter, *Gene,* 5, 59, 1979.

104. **Roberts, T. M., Bikel, I., Rogers-Yocum, R., Livingston, D. M., and Ptashne, M.,** Synthesis of simian virus 40 t antigen in *E. coli, Proc. Natl. Acad. Sci.,* 76, 5596, 1979c.

105. **Seeburg, P. H., Shine, J., Marshall, J. A., Baxter, J. D., and Goodman, H. M.,** Nucleotide sequence and amplification in bacteria of structural gene for rat growth hormone, *Nature,* 220, 486, 1977.

106. **Broome, S. and Gilbert, W.,** Immunological screening method to detect specific translation products, *Proc. Natl. Acad. Sci.,* 75, 2746, 1978.

107. **Talmadge, K., Brosius, J., and Gilbert, W.,** An ''internal'' signal sequence directs secretion and processing of proinsulin in bacteria, *Nature,* 294, 176, 1981.

108. **Uhlin, B. E., Molin, S., Gustafsson, P., and Nordstrom, K.,** Plasmids with temperature-dependent copy number for amplification of cloned genes and their products, *Gene,* 6, 91, 1979.

109. **Brent, R. and Ptashne, P.,** Mechanism of action of the *lexA* gene product, *Proc. Natl. Acad. Sci.,* 78, 4204, 1981.

110. **Wu, R., Grossman, L., and Moldane, K.,** Recombinant DNA, Vol. 101 Part C, *Methods in Enzymology,* Academic Press, New York, 1983.

111. **Setlow, J. and Hollaender, A.,** *Genetic Engineering,* Vol. 4, Plenum Press, 1982.

112. **Williamson, R.,** *Genetic Engineering,* Vol. 4, Academic Press, 1983.

113. **Harris, T. J. R.,** Expression of eukaryotic genes in *E. coli,* in *Genetic Engineering,* Williamson, R., Ed., Academic Press, 1983, 127.

114. **Southern, E.,** Detection of specific sequences among DNA fragments separated by gel electrophoresis, *J. Mol. Biol.,* 98, 503, 1975.

115. **Bollon, A. P., Fuke, M., and Torczynski, R. M.,** A procedure for isolation of α-interferon genes with oligoneucleotide probes, in *Methods in Enzymology, Interferons C,* Pestka, S., Ed., Academic Press N.Y., in press.

Chapter 2

FROM SOMATOSTATIN TO HUMAN INSULIN

Arthur D. Riggs, Keiichi Itakura, and Herbert W. Boyer

TABLE OF CONTENTS

I. INTRODUCTION

On October 29, 1982, the United States Food and Drug Administration gave marketing approval for Humalin®, a human insulin made by bacteria. A month earlier, Humalin had been approved for sale in the United Kingdom. These approvals represent an important milestone, because human insulin thereby became the first clinically significant product resulting directly from the new recombinant DNA technologies. In this review, we will describe the research done at the City of Hope Research Institute, the University of California at San Francisco, and Genentech Inc., that demonstrated scientific feasibility and led directly to the construction of the bacterial strains now being used for insulin production by the Eli Lilly Company.

A common misconception is that human genes for human insulin were placed in bacteria and made to function. This was not the case for either our insulin work[1] or our earlier work on somatostatin.[2] Instead, completely artificial genes were designed and synthesized, using codons preferred by *Escherichia coli*. Thus, man-made genes, designed to function in a bacterium, were used to direct the production of the desired peptide hormones. When we began the work, it was not at all certain that we knew enough about genes and the details of bacterial gene expression to be successful. However, it is now clear that the accumulated knowledge was adequate, and the time was right for the design and chemical synthesis of genes for small hormones.

In 1976, Robert Swanson and Herbert Boyer formed a company called Genentech Inc. Their aim was to raise venture capital and use recombinant DNA and chemical DNA synthesis techniques to engineer bacterial cells to produce useful products, such as human hormones. The first contract let by Genentech was to Itakura and Riggs, for the chemical synthesis of somatostatin, a mammalian hormone containing only 14 amino acids. Somatostatin was chosen for the pilot study because of its small size and the availability of good radioimmunoassays for its detection. It is interesting that attempts had been made to obtain grants for the somatostatin work, but these were not funded because of such criticism as "It seems like just an intellectual exercise....somatostatin is not likely to be of great clinical significance." In retrospect, it is evident that the somatostatin project was one intellectual exercise that had very significant and rapid practical applications, leading directly to the bacterial synthesis of human insulin,[1] growth hormone,[3] and interferon,[4] and catalyzing the growth of the new biotechnology industry.

II. CHEMICAL DNA SYNTHESIS

Because of its central importance to the somatostatin and insulin projects, as well as to many subsequent genetic engineering projects (e.g., growth hormone and interferon), we will first describe the triester method of chemical DNA synthesis (see Figure 1) as it was used for the synthesis of the somatostatin and insulin genes.[2,5]

Starting with nucleosides, a library of fully protected triester trimers was made. The trimers were condensed to yield, for example, dodecamers, still in fully protected triester form. All protecting groups were then removed by treatment with acetic acid and NH_4OH, generating the desired water-soluble single-stranded DNA fragment. The last step was a careful purification of the DNA fragments by high performance liquid chromatography. As we will describe later, the oligonucleotides were assembled enzymatically to form a DNA duplex, which was then cloned.

Improvements in the DNA synthesis technology have been dramatic over the last few years and the synthesis of oligodeoxyribonucleotides is now highly automated and rapid; a heptadecamer can easily be made overnight using solid-phase methodology,[6] and chemical DNA synthesis is no longer the rate-limiting step for most projects.

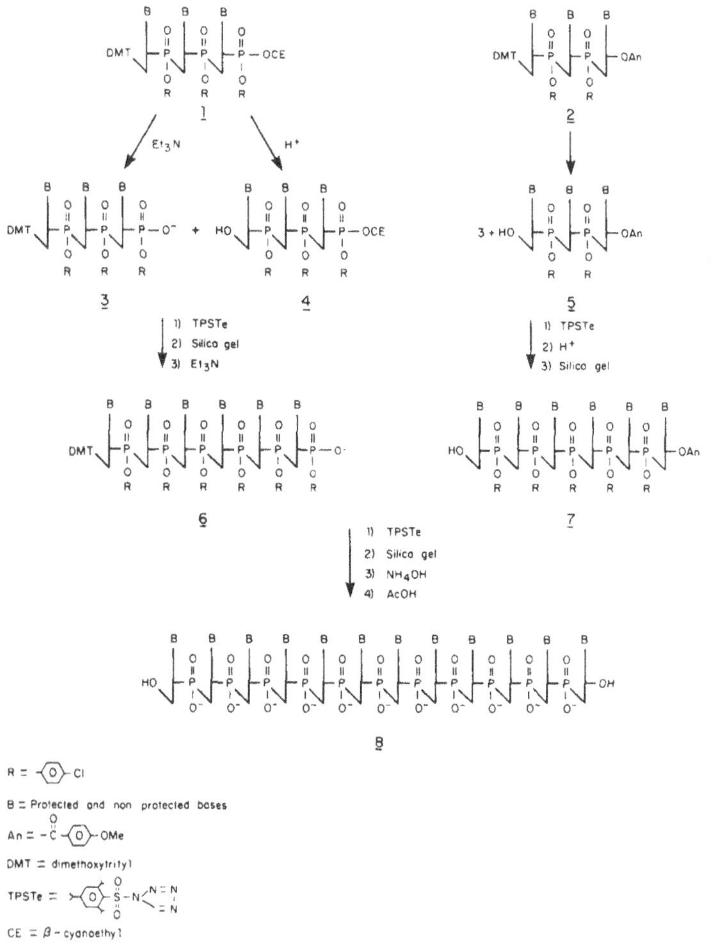

FIGURE 1. The chemical synthesis of oligodeoxyribonucleotides for somatostatin and insulin genes. The basic units used to construct polynucleotides are two types of trimer block, the bifunctional trimer 1 and the 3'-terminus trimer 2. The bifunctional trimer 1 was hydrolyzed to the 3'-hydroxylphosphodiester component 3 with pyridine/triethylamine/water (3:1:1 vol/vol) and also to the 5'-hydroxyl component 4 with 2% benzenesulfonic acid. The 3'-terminus block 2, protected by an anisoyl group at 3'-hydroxy, was treated with 2% benzenesulfonic acid to give the 5'hydroxyl block 5. The coupling reaction of an excess of the 3'-phosphodiester trimer 3 (1.5 M equivalent) with the 5'-hydroxyl component 4 or 5 (1 M equivalent) in the presence of 2,4,6-triisopropylbenzenesulfonyl tetrazolide (TPSTe, 3-4 equivalents) went almost to completion in 3 hr. To remove the excess of the 3'-phosphodiester block 3, the reaction mixture was passed through a short silica gel column set up on a sintered glass filter. The column was washed, first with CHCl₃ to elute some side products and the coupling reagent, and then with CHCl₃/MeOH (95:5 vol/vol) in which almost all of the fully protected oligomer was eluted. Under these conditions, the charged compound 3 remained in the column. Similarly, block couplings were repeated until the desired length was constructed. (From Crea, R., Kraszewski, A., Hirose, T., and Itakura, K., *Proc. Natl. Acad. Sci. U.S.A.*, 75, 5765, 1978. With permission.)

III. SOMATOSTATIN

In 1977, a chemically synthesized somatostatin gene was made, introduced into bacteria, and shown to function.[2] As outlined in Figure 2, DNA fragments for a 57-bp somatostatin gene were synthesized by the triester method, joined by ligase to form double-stranded DNA,

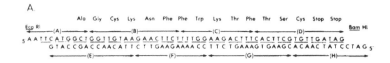

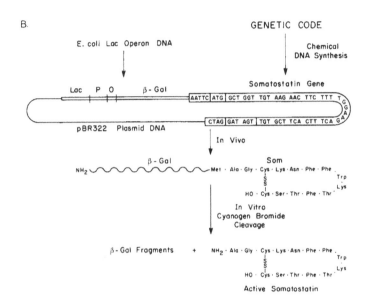

FIGURE 2.   (A) The chemically synthesized somatostatin gene. Eight oligodeoxy-
nucleotides (fragments A to H) were synthesized by the modified triester method.[15] The
codons are underlined and their corresponding amino acids are given. The 8 fragments
were designed to have at least 5 nucleotide complementary overlaps to ensure correct
joining by T4 DNA ligase to give duplex DNA 60 nucleotides long. (B) Schematic
overview of the somatostatin project. The gene for somatostatin was fused to the *E.
coli* β-galactosidase gene on the plasmid pBR322. After transformation into *E. coli*,
the plasmid directs the synthesis of a chimeric protein that has somatostatin on a short
tail on the end of β-galactosidase. Somatostatin is clipped off by treatment with cyanogen
bromide, which cleaves specifically at methionine residues. (Reprinted with permission
from Itakura, K., Hirose, T., Crea, R., Riggs, A. D., Heyneker, H. L., Boliver, F.,
and Boyer, H. W., *Science*, 198, 1056, 1977. Copyright 1977 by the AAAS.)

and inserted near the end of the bacterial gene for beta-galactosidase. In vivo, somatostatin
was made as a short tail on the end of the beta-galactosidase polypeptide chain. Somatostatin
was cleaved from the fused protein product by treatment in vitro with cyanogen bromide,
which very specifically and efficiently cleaves polypeptide chains at methionine. Because
the gene was made synthetically, it was easy to arrange for a methionine to be at the junction
of beta-galactosidase and somatostatin. After treatment with cyanogen bromide, somatostatin
radioimmune activity was readily detected. The purified hormone product was shown later
to be completely identical to natural mammalian somatostatin.

IV. INSULIN

As soon as the results of the stomatostatin project demonstrated the feasibility of the
approach, work began on two genes for the A and B chains of human insulin.[5] It was
necessary to make 29 oligonucleotides, which were assembled and joined by ligation to
make a total of 181 base pairs of duplex DNA (Figure 3). Figures 4 and 5 illustrate how
the insulin A-chain was assembled, cloned, and positioned at the end of beta-galactosidase.

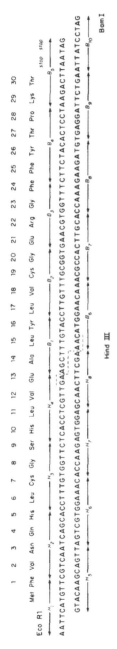

FIGURE 3. Chemically synthesized human insulin genes. The genes for human insulin A and B chains were designed from the amino acid sequences of the human polypeptides. The 5′ ends of each gene have single-stranded cohesive termini for the *Eco* RI and *Bam* I restriction endonucleases for correct insertion of each gene into plasmid pBR322. A *Hind* III endonuclease recognition site was incorporated into the middle of the B chain gene for the amino acid sequence *Glu-Ala* to allow amplification and verification of each half of the gene separately before construction of the whole B-chain gene. The B- and the A-chain genes were designed to be built from 29 different oligodeoxyribonucleotides, varying from decamers to pentadecamers. Each arrow indicates the fragment synthesized by the improved phosphotriester method, H1 to H8 and B1 to B12 for the B-chain gene, and A1 to A11 for the A-chain gene. (From Crea, R., Kraszewski, A., Hirose, T., and Itakura, K., *Proc. Natl. Acad. Sci. U.S.A.*, 75, 5765, 1978. With permission.)

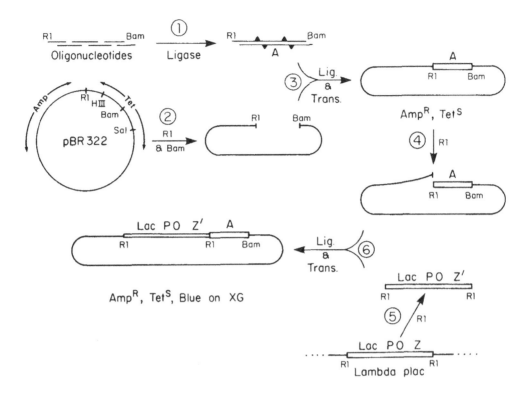

FIGURE 4.   The construction of a plasmid DNA containing a synthetic insulin A-chain gene inserted at the end of a β-galactosidase gene. The procedures are described in the text, and details are given in Goeddel et al.[1] Only an explanation of the symbols is given here. The symbol A represents the synthetic A-chain gene. pBR322 is a well-characterized plasmid cloning vector containing two antibiotic resistance genes, ampicillin (*Amp*) and tetracycline (*Tet*), and several convenient restriction endonuclease sites including *Eco* RI (R1) and *Bam* HI (*Bam*). Lambda plac is a lambda transducing phage carrying the entire *E. coli* operon, which includes the lac promoter (P), the lac operator (O), and the entire β-galactosidase structural gene (Z). There is an *Eco* RI endonuclease site to the left of the operon and also one near the end of the β-galactosidase gene; thus, the lac operon DNA fragment can be readily obtained. The phenotype of the bacterial strains successfully infected with the desired plasmid is shown. For example, the A-chain-producing strains would be ampicillin resistant (Amp^R), tetracycline sensitive (Tet^S), and the colonies would be blue on a special indicator agar called Xg. (From Riggs, A. D. and Itakura, K., *Am. J. Hum. Genet.*, 31, 531, 1979. With permission of The University of Chicago Press.)

Step 1 (shown in Figure 4) was joining the small oligonucleotides. Because they were designed to have complementary overlaps, they assembled themselves and were joined to give duplex DNA by the action of the T4 DNA ligase. The gene was designed to have restriction enzyme sites at each end (*Eco* RI on the left and *Bam*HI on the right). Step 2 was preparation of the plasmid DNA cloning vector pBR322. Preparation included treatment with *Eco*RI and *Bam*HI restriction enzymes, which cut out a small piece of the plasmid and provided a site for insertion of the synthetic A gene. In step 3, the prepared plasmid and synthetic DNA were mixed, joined by T4 ligase, and cloned by transformation of *E. coli*. A clone was obtained that contained a correct insulin A gene, as verified by direct DNA sequencing. Next, a DNA fragment containing most of the *E. coli* lac operon, including the lac promoter, operator, and the first 1006 amino acid codons of beta-galactosidase, was inserted (steps 4, 5, and 6). This led to a clone making insulin-beta-galactosidase-fused protein.[1]

Approximately 20% of the bacterial protein produced by the insulin strains was beta-galactosidase-insulin-fused protein (Figure 5). The fused protein is insoluble and was enriched to more than 50% purity by low speed centrifugation. The insulin chains then were

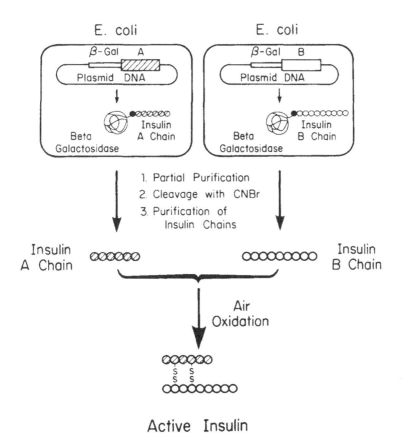

FIGURE 5. Schematic overview of the strains and procedures for the production of human insulin by bacteria. Two *E. coli* strains were constructed having chemically synthesized insulin A- or B-chain genes inserted into the β-galactosidase gene (β-gal) of a plasmid cloning vector. In vivo, a fused protein is made, mostly β-galactosidase, but with an insulin tail joined by a methionine. In vitro, the insulin peptide chain is clipped off by treatment with cyanogen bromide. After separate purification, the insulin A and B chains are joined by air oxidation. (From Riggs, A. D. and Itakura, K., *Am. J. Hum. Genet.*, 31, 531, 1979. With permission of The University of Chicago Press.)

cleaved from the fused protein by overnight treatment with cyanogen bromide in 70% formic acid at room temperature. The free insulin chains were converted to S-sulfonated derivatives and purified by ion exchange chromatography, gel filtration, and reverse phase high performance chromatography. In our published work,[1] the B-chain was not purified to homogeneity. However, this now has been done. No major problems were encountered.

Sulfonated derivatives of the insulin chains were made before purification to ensure stability by preventing the premature formation of disulfide bonds. Fortunately, the best method for joining chains[7] starts with the S-sulfonated derivatives. With a fivefold excess of A-chain, up to 80% correct joining of the B-chain to the A-chain can be obtained. To detect insulin chains produced in the bacteria, we adapted the Katsoyannis reconstitution procedure[7] to the microgram scale. Using the procedure outlined in Figure 6, we had no difficulty joining chains to produce radioimmune active insulin in control experiments and with partially purified bacterial insulin chains.[1] The microreconstitution assay was used to follow the individual insulin chains during purification. In our laboratory-scale experiments[1] we obtained approximately 1 mg of insulin chain per liter of culture. These yields were high enough to stimulate large-scale commercial development work (not described here), which has led to higher yields.

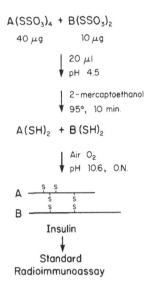

FIGURE 6.   Insulin chain joining according to the Katsoyannis procedure. S-sulfonated B chain is mixed with a 4-fold excess of S-sulfonated A chain, and then free sulfhydryls are produced by reduction with mercaptoethanol. After removal of the excess mercaptoethanol by extraction with ethyl acetate, the pH is adjusted to 10.6 and the chains are joined by air oxidation at room temperature overnight (O.N.). A standard radioimmunoassay (Pharmacia) was used to measure radioimmune insulin. (From Riggs, A. K., Itakura, K., Crea, R., Hirose, T., Kraszewski, A., Goeddel, D., Kleid, D., Yansura, D. G., Bolivar, F., and Heyneker, H. L., *Recent Prog. Horm. Res.*, 36, 261, 1980. With permission.)

V. LARGE-SCALE PRODUCTION AND CLINICAL TRIALS

The laboratory demonstration of the bacterial production of human insulin was only the beginning of the difficult process of bringing human insulin to the public. Most of the authors of this paper were not involved in scale-up and product development, and much credit should be given to numerous investigators at Genentech and also at Eli Lilly & Co., to whom Genentech sold manufacturing and marketing rights. No insurmountable technical obstacles were encountered in the scale-up. In fact, the joining of the two chains, being a bimolecular process, has proved to be more efficient on a large scale. Eli Lilly & Co. launched a major effort to bring the human insulin to the public by 1982 and they met their goal. All chemical tests and clinical trials have been favorable,[8-13] and sales have begun.

REFERENCES

1. Goeddel, D. V., Kleid, D. G., Bolivar, F., Heyneker, H. L., Yansura, D. G., Crea, R., Hirose, T., Kraszewski, A., Itakura, K., ånd Riggs, A. D., Expression in *Escherichia coli* of chemically synthesized genes for human insulin, *Proc. Natl. Acad. Sci. U.S.A.*, 76, 106, 1979.
2. Itakura, K., Hirose, T., Crea, R., Riggs, A. D., Heyneker, H. L., Boliver, F., and Boyer, H. W., Expression in *Escherichia coli* of a chemically synthesized gene for the hormone somatostatin, *Science*, 198, 1056, 1977.
3. Goeddel, D. V., Heyneker, H. L., Hozumi, T., Arentzen, R., Itakura, K., Yansura, D. G., Ross, M. J., Miozzari, G., Crea, R., and Seeburg, P. H., Direct expression in *Escherichia coli* of a DNA sequence coding for human growth hormone, *Nature*, 281, 544, 1979.
4. Goeddel, D. V., Yelverton, E., Ullrich, A., Heyneker, H. L., Miozzari, G., Holmes, W., Seeburg, P. H., Dull, T., May, L., Stebbing, N., Crea, R., Maeda, S., McCandliss, R., Sloma, A., Tabor, J. M., Gross, M., Familletti, P. C., and Pestka, S., Human leukocyte interferon produced by *E. coli* is biologically active, *Nature*, 287, 411, 1980.
5. Crea, R., Kraszewski, A., Hirose, T., and Itakura, K., Chemical synthesis of genes for human insulin, *Proc. Natl. Acad. Sci. U.S.A.*, 75, 5765, 1978.
6. Tan, Z-K., Ikuta, S., Huang, T., Dugaiczyk, A., and Itakura, K., Solid-phase synthesis of polynucleotides. VII. The phosphotriester solid-phase method, *Cold Spring Harbor Symp. Quant. Biol.*, 47, 1982, in press.
7. Katsoyannis, P. G., Trakatellis, A. C., Johnson, S., Zalut, C., and Schwartz, G., Studies on the synthesis of insulin from natural and synthetic A and B chains. II. Isolation of insulin from recombination mixtures of natural A and B chains, *Biochemistry*, 6, 2642, 1967.
8. Keen, H., Pickup, J. C., Bilous, R. W., Glynne, A., Viberti, G. D., Jarrett, R. J., and Marsden, R., Human insulin produced by recombinant DNA technology: safety and hypoglycaemic potency in healthy men, *Lancet*, 2, 39, 1980.
9. Baker, R. S., Ross, J. W., Schmidtke, J. R., and Smithe, W. C., Preliminary studies on the immunogenicity and amount of *Escherichia coli* polypeptides in biosynthetic human insulin produced by recombinant DNA technology, *Lancet*, 2, 1139, 1981.
10. Botterman, P., Gyaram, H., Wahl, K., Ermler, R., and Lebender, A., Pharmacokinetics of biosynthetic human insulin and characteristics of its effect, *Diabetes Care*, 4, 168, 1981.
11. Federlin, K., Laube, H., and Velcovsky, H. G., Biologic and immunologic in vivo and in vitro studies with biosynthetic human insulin, *Diabetes Care*, 4, 170, 1981.
12. Keefer, L. M., Piron, M.-A., and De Meyts, P., Human insulin prepared by recombinant DNA techniques and native human insulin inateract identically with insulin receptors, *Proc. Natl. Acad. Sci., U.S.A.*, 78, 1391, 1981.
13. Clark, A. J. L., Knight, G., Wiles, P. G., Keen, H., Ward, J. D., Cauldwell, J. M., Adeniyi-Jones, R. O., Leiper, J. M., Jones, R. H., MacCuish, A. C., Watkins, P. J., Glynne, A., and Scotton, J. B., Biosynthetic human insulin in the treatment of diabetes, *Lancet*, 2, 354, 1982.
14. Riggs, A. D. and Itakura, K., Synthetic DNA and medicine, *Am. J. Hum. Genet.*, 31, 531, 1979.
15. Itakura, K., Katagiri, N., Narang, S. A., Bahl, C. P., Marians, K. J., and Wu, R., Chemical synthesis of the lactose operator of *E. coli*, *J. Biol. Chem.*, Vol. 250, 4592, 1975.
16. Riggs, A. D., Itakura, K., Crea, R., Hirose, T., Kraszewski, A., Goeddel, D., Kleid, D., Yansura, D. G., Bolivar, F., and Heyneker, H. L., Synthesis, cloning, and expression of hormone genes in *Escherichia coli*, *Recent Prog. Horm. Res.*, 36, 261, 1980.

Chapter 3

YEAST: AN ALTERNATIVE ORGANISM FOR FOREIGN PROTEIN PRODUCTION

Ronald A. Hitzeman, Christina Y. Chen, Frank E. Hagie, June M. Lugovoy, and Arjun Singh

TABLE OF CONTENTS

I. INTRODUCTION

Recombinant DNA technology used for the production of pharmaceutically useful poly-peptides such as insulin[1] human growth hormone,[2] and the interferons[3-7] has thus far mainly been focused on *E. coli* expression systems. However, within the last few years alternative expression systems such as yeast,[8] tissue culture,[9] *Bacillus subtilis,*[10] *Pseudomonas,*[11] and *Streptomyces*[12] have also attracted the interest of applied scientists. Since these systems are relatively new and have not been characterized extensively, a great deal of basic research examining gene expression and regulation is needed before these organisms can be harnessed for the recombinant DNA industry. The focus of this report will be on one of these alternative organisms, the yeast *Saccharomyces cerevisiae*, and the progress that has been made by us and various investigators in the production of foreign (heterologous) gene products by this organism.

II. ADVENT OF YEAST MOLECULAR BIOLOGY

Recombinant DNA research in yeast became possible in 1978 when Hinnen et al.[13] demonstrated the first successful introduction (transformation) of foreign DNA into yeast. This was done by enzymatic removal of the yeast cell wall (generation of spheroplasts) and incubation of plasmid DNA with the spheroplasts in the presence of calcium ions and polyethylene glycol. The entry of the plasmid DNA into the cell was detected by comple-mentation of *leu2⁻* yeast with the yeast *LEU2* gene, which had previously been isolated by Ratzkin and Carbon.[14] This initial procedure resulted in integration of the plasmid DNA into the yeast genome; however, similar procedures have also been used by Beggs,[15] Struhl et al.,[16] and Hsiao and Carbon[17] to obtain transformation with autonomously replicating yeast plasmids containing yeast 2 μm plasmid[15] or chromosomal origins of replication.[16,17] This procedure, along with the ability to reisolate plasmid DNA from yeast extracts by transfor-mation of *E. coli,*[15] has resulted in the isolation of many genes from yeast. Among the genes isolated and characterized by this methodology are the *HIS4,*[18] isocytochrome c,[19] and the alcohol dehydrogenase I[20] and II[21] genes.

III. EXPRESSION OF HETEROLOGOUS GENE PRODUCTS IN YEAST

The first successful attempts to express heterologous (non-yeast) genes in yeast were performed using *E. coli* genes. Hollenberg[22] as well as Chevallier and Aigle[23] demonstrated that the β-lactamase gene of *E. coli* plasmid pBR322 was expressed in yeast, using tran-scriptional control regions in the *E. coli* DNA.[24] Hollenberg[22] and Cohen et al.[25] showed that the chloramphenicol acetyltransferase gene from plasmid pBR325 was functionally expressed in yeast.[25]

In addition to *E. coli* genes, expression of some higher eukaryotic genes has also been examined. Beggs et al.[26] demonstrated that the genomic rabbit β-globin gene in yeast produced a shortened mRNA of insufficient length to produce β-globin. Whether yeast used the rabbit promoter, a fortuitous promoter region in the rabbit DNA, or a yeast promoter was not determined. However, transcription appeared to begin near its normal initiation site in the rabbit DNA and ended prematurely or was cleaved in the second intron of the gene. Thus no normal β-globin protein could be made. Later the *ADE8* gene from *Drosophila melanogaster* was expressed in yeast by complementing the corresponding yeast mutation as described by Henikoff et al.[28] This gene was probably expressed using fortuitous tran-scription signals in the *Drosophila* DNA (B. Hall, personal communication).[81]

Fraser and Bruce[27] used the cloned cDNA of ovalbumin as a model system in an attempt to avoid the problems with introns suggested by the experiments of Beggs et al.[26] mentioned above. They placed the cDNA in a yeast plasmid with pBR322 DNA flanking the ovalbumin gene. Although the exact nature of the product was not thoroughly characterized, an ovalbumin product (1500 mol/cell) was detected by a radioimmune assay. They demonstrated that the transcription start for the mRNA in this system probably originated in the pBR322 DNA instead of in yeast DNA as they had expected.

In order to obtain efficient expression of a heterologous gene in yeast, we decided to use yeast DNA transcriptional signals adjacent to the gene on a plasmid capable of replication in yeast. The human leukocyte interferon D (IFN-α1 or LeIFD) gene or cDNA(s) was chosen for these experiments as previously described.[8] The promoter (transcription initiation) fragment was obtained from the 5'-flanking sequence of the yeast alcohol dehydrogenase I *(ADH1)* gene[20,29] and the terminator fragment (transcription termination and/or polyadenylation signals) was obtained from the yeast *TRP1* gene[30,31] 3'-flanking region. Messenger RNA size analysis demonstrated that the transcript was of the appropriate length to have initiated immediately before the IFN-α1 fragment and have ended in the 3'-flanking region of the *TRP1* gene. The expression levels of the IFN reached 0.04 to 0.2% of total yeast protein when a plasmid containing the *ars1* origin of replication[30] was used and about 1% in a plasmid containing the yeast 2 μm origin of replication. The product also had biological activity in an antiviral bioassay[32] and was identical with IFN-α1 obtained from human cells except for the amino-terminal methionine.[8] The amino-terminal methionine had been added to allow for translation initiation after removal of the secretion signal pre-sequence from the IFN cDNA and replacement by an ATG-translational start as shown in Figure 1 (construction B converted to A). Unlike some yeast genes that have been expressed in *E. coli*,[33] this efficient yeast expression system produced no detectable IFN-α1 product in *E. coli*, showing that not all yeast transcriptional or translational controls are functional in *E. coli*.

The 5'-flanking sequence of the *ADH1* gene was used because of the highly expressed nature of the gene (single copy per haploid cell produces about 1% of total protein) and the known DNA sequence.[29] Furthermore, Ammerer et al.[34] had shown that the *ADH1* 5'-flanking sequence was capable of conferring its transcription starting specificity when fused to the yeast *CYC1* coding sequence. Several promoter fragments were made from the 5'-flanking sequence of the *ADH1* gene by *Bal*31 digestion from a restriction site in the structural gene through the ATG-translational start. It was determined that up to 32 bp could be removed upstream of the ATG without grossly affecting expression of IFN-α1, suggesting that this region is not critical to either translation or transcription. Interestingly, this same promoter system efficiently transcribed the rat growth hormone cDNA but no translation product was observed.[34] This may have been due to junction effects (DNA sequence between promoter and structural gene) on translation initiation, codon bias effects on translation efficiency,[35] or extreme instability of the protein product.

The same *ADH1* promoter fragment was recently used by Valenzuela et al.[36] to express the surface antigen (HBsAg) from hepatitis B virus. Expression was sufficient to produce about 10 to 25 μg/ℓ (0.01 to 0.025 percent of cellular protein) of a particle form of HBsAg very similar to the natural serum-derived 22 nm particle. This similarity was demonstrated by sucrose gradient sedimentation and equilibrium density centrifugation, as well as by electron microscopic comparisons. Miyanohara et al.[37] demonstrated the formation of a 20 to 22 nm particles in yeast using the 5'-flanking sequence of the yeast repressible acid phosphatase gene[38] and showed derepression of HBsAg synthesis when the yeast were grown in phosphate-free medium. We have also expressed the HBsAg[39] using a promoter fragment from the yeast 3-phosphoglycerate kinase gene.[40,41] We found a particle form (0.05% of yeast protein), but after detergent lysis of cells observed 20 to 40 times this amount as a monomer form (comprising 1 to 2% of the total cellular protein). The large amount of monomer in comparison with particle suggests that a vaccine may be more efficiently obtained by first purifying the monomer, followed by reformation of a particle form.[36,37,39]

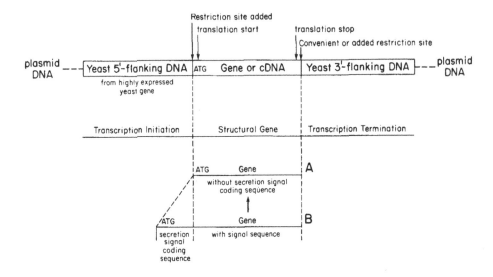

FIGURE 1.   Heterologous gene expression system.

IV. CONSTRUCTION OF A PORTABLE PROMOTER FRAGMENT

As mentioned above, three different yeast promoter fragments have been utilized in the expression of heterologous genes in yeast. We now present in more detail the procedures used to construct one such fragment. Figure 2 illustrates a partial restriction map of a cloned 3.1 kbp *Hind*III fragment from chromosome III of yeast, which contains the gene for yeast 3-phosphoglycerate kinase (PGK).[40] The exact location of the PGK structural gene was determined by DNA sequencing from the *Pvu*I site to the *Hind*III site on the right side of Figure 2.[41] This region has been identified as the start of the structural gene because of the length of the open reading frame that follows (matching with PGK polypeptide size), the DNA sequence characteristics 5′ to the ATG (see Reference 34), and 65% homology of yeast PGK protein sequence (derived from DNA sequence) with human[43] and horse[44] PGK sequences.

Given the start of the structural gene, we needed a design for the isolation of the 5′-flanking sequence of the PGK gene on a specific restriction fragment to be used as a portable promoter fragment for expression of heterologous genes. For the previously mentioned *ADH1* gene this was done by *Bal*31 digestion through the ATG into the 5′-flanking sequence followed by the addition of *Eco*RI linkers.[8,34] Since deletion of up to 32 base pairs of 5′-flanking sequence adjacent to the ATG had no gross effects on the expression of a heterologous gene (leukocyte interferon D), *Bal*31 digestions were not thought to be necessary for construction of the PGK promoter fragment.

Instead, we used the primer repair technique described by Goeddel et al.[6] An oligonucleotide (12 nucleotides long) was synthesized which was complementary to the −10 through −21 positions of the nucleotide sequence shown in the insert of Figure 2. When this oligonucleotide was hybridized with denatured DNA from this region of the PGK gene in the presence of Klenow DNA polymerase I and deoxyribonucleotide triphosphates, polymerization occurred 5′→3′ through the upstream *Sau*3A site while the 3′→5′ exonuclease degraded phosphodiester bonds from the 3′ end of the fragment (*Hinc*II site in the PGK gene) until the double stranded region was reached. By cutting the repair product with *Sau*3A, a small region of the PGK promoter fragment was isolated as a *Sau*3A end to blunt end DNA (−46 to −10) fragment. The loss of −9 through −1 of the 5′-flanking sequence was thought not to be a problem in view of the *ADH1* promoter fragment results already discussed.

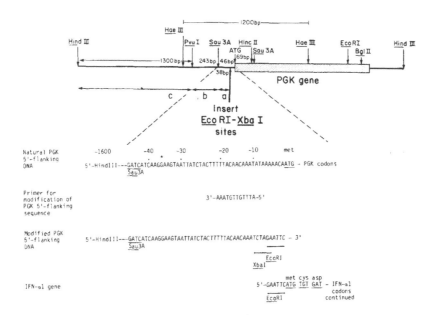

FIGURE 2. Partial restriction and sequence map of the PGK gene. The PGK structural gene is shown as a bar region in the partial restriction map of a 3.1 *Hind*III fragment. Note that distances are not drawn to scale; however, distances of fragments relevant to the discussed constructions are designated. The inserted 5' to 3' sequence (left to right) is from the *Sau*3A site through the ATG of coding sequence. A complementary oligonucleotide (12 nucleotides) for the primer repair reaction (6) was made as shown. Using this primer the modified PGK 5'-flanking sequence was constructed for convenient attachment of gene such as mature IFN-αl as shown.

The fragment which contained a blunt end and a *Sau*3A end (sequence "a" in Figure 2) was then ligated into a plasmid vector containing a *Bam*HI sticky end and a filled-in *Xba*I blunt end. This allowed a portion of the 5'-flanking sequence of the PGK gene to be isolated from the vector as a *Sau*3A to *Xba*I restriction fragment. However, since this fragment was small and since transcriptional starting efficiency and control of yeast genes is sometimes affected by yeast DNA far upstream from the ATG-translational start,[19,45-47] the plan to make a promoter fragment required the reconstruction of 5'-flanking sequence upstream from the *Sau*3A site to the *Hind*III site (thus sequences "b" and "c" in Figure 2 were added back to sequence "a"). During this reconstruction an *Eco*RI site was added adjacent to the *Xba*I site so that the 1600 bp promoter fragment could be isolated as a *Hind*III-to-*Xba*I or as a *Hind*III-to-*Eco*RI fragment (Figure 2, bottom).

V. CONSTRUCTION OF A HETEROLOGOUS GENE EXPRESSION PLASMID USING THE PGK PROMOTER

The plasmid YEp1PT in Figure 3 contains the PGK promoter fragment described above and other expression, selection, and maintenance components for expression of inserted heterologous genes in a unique *Eco*RI site. This plasmid includes the large *Eco*RI to *Bam*HI fragment of pBR322[48] containing the ampicillin resistance (Ap[R]) gene and replication origin for selection and stable growth in *E. coli*. YEp1PT also contains the yeast *TRP1* gene on an *Eco*RI to *Pst*1 DNA fragment originating from yeast chromosome IV.[30,31] This gene permits selection for the plasmid in trp1⁻ yeast growing in medium lacking tryptophan. The other origin of replication in YEp1PT is contained on a 2.0 kbp fragment (*Eco*RI to *Pst*1) from the endogenous yeast 2 μm plasmid[49,50] and allows autonomous replication and maintenance at high copy number in 90 to 95% of yeast cells grown selectively.

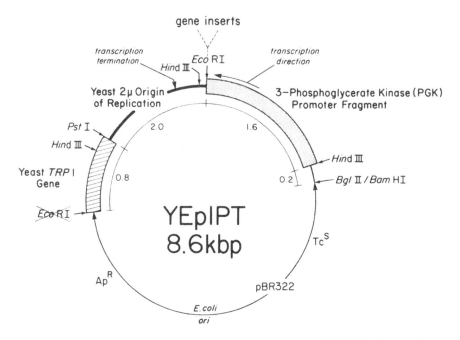

FIGURE 3.   Yeast expression plasmid. The partial restriction map of YEp1PT is shown with components designated that are required for transcription and translation of a heterologous gene inserted at the single *Eco*RI site.

The yeast PGK promoter fragment (*Hind*III to *Eco*RI) initiates transcription near the only *Eco*RI restriction site in the plasmid (the other *Eco*RI site was removed as previously described by Hitzeman et al.[8]) which proceeds in the direction indicated (Figure 3). As shown in Figure 2, the sequence 5'-TCTAGAATTC-3' which contains *Eco*RI and *Xba*I restriction sites for convenient attachment of heterologous genes, has been substituted for the 10 bp of normal sequence preceding the ATG translation initiation codon of PGK. A *Hind*III/*Bgl*II fragment from the yeast *TRP1* gene region[30,31] was used as a *Hind*III to *Bgl*II converter for ligation with the *Bam*HI site of pBR322. The 2.0 kbp fragment from 2 μm plasmid DNA also contains the transcription termination and polyadenylation signals which are normally the signals for the "Able" (or *FLP*) gene found in 2 μm plasmid.[50] As discussed below, such a region appears to be essential for heterologous gene expression in yeast. When heterologous genes such as leukocyte interferons IFN-α1, IFN-α2, and hepatitis surface antigen are inserted at this *Eco*RI site in the correct orientation, they are expressed at levels of 1 to 2% of total yeast protein.

Comparisons have been made of this promoter fragment with the *ADH1* promoter fragment (fragment 906,[8]) already discussed. In plasmids containing either *ars1*[30] or 2 μm[49] origins of replication, both promoter fragments express essentially identical levels of IFN-α1. These results suggest that the two promoters are comparable in transcription initiation and thus equally useful for expression of heterologous genes in yeast.

VI. CHROMOSOMAL INTEGRATION OF HETEROLOGOUS GENE EXPRESSION SYSTEMS

Expression of several foreign genes in yeast has been achieved using autonomously replicating hybrid plasmids as described above, however, it is also possible to place such expression systems into the yeast genome. As an example of such a system, we have modified the initial IFN-α1 expression plasmid pFRS36[8] by removing the yeast origin of replication, *ars1*,[30] so that the plasmid must integrate into a yeast chromosome for maintenance and expression. To remove the *ars1* function, DNA sequences distal to the *Pst*I site in the 1.45 kb fragment[30,31] containing the *TRP1* gene were removed. The resulting plasmid, pAS1 (not shown), which contains only 51 nucleotides of yeast DNA following the termination codon of the *TRP1* gene, is not capable of replication in yeast although the *TRP1* gene is still functional.

Yeast strain RH218[42] was transformed with the plasmid, and the Trp$^+$ transformants obtained were assayed for IFN-α1 activity. Interferon activity in over 100 independent transformants varied from undetectable levels to about 0.2% of total yeast protein at Abs^{660} of 1.0. When several transformants were checked for stability of Trp$^+$ phenotype after growth in both selective and non-selective media, all were completely stable. We have not determined the chromosomal location of the integration events. However, genetic analysis of hybrids generated from the integrants has shown that in none of the 4 transformants tested had the pAS1 plasmid integrated at the *TRP1* locus.

We have checked whether the expression levels correspond to the number of integrated copies of the plasmid. Total DNA, prepared from RH218 and three transformants producing various levels of interferon D, was digested with *Bgl*II, subjected to electrophoresis, transferred to nitrocellulose, and probed with the IFN-α-1-*TRP1* junction fragment. The *Bgl*II-*Xba*I junction fragment spans 360 bp of the IFN-α1 gene and 190 bp from the *TRP1* fragment. As seen in Figure 4, the level of hybridization is in general agreement with the IFN-α1 expression levels. No additional hybridizing DNA was detected from transformant 1 (lane 1) which did not produce any detectable interferon. The bands A and B in transformants 2 and 3 apparently resulted from hybridization of the probe with a DNA fragment spanning the site of integration. It should be noted that the integrative plasmid used contained only one *Bgl*II site (located in IFN-α1 gene). Band B suggests that multiple copies of the plasmid have integrated in a tandem array in the transformant 3.

We have also constructed linear integrative expression units which contain the PGK promoter, HBsAg gene, *TRP1* terminator and the *LEU2* gene. These transformants produce varying levels of hepatitis surface antigen as determined by radioimmunoassays. By Southern analysis we have found that, like the IFN-α1 plasmids, the expression levels correspond to the number of integrated expression units with up to 10 to 15 direct repeats of this unit. Such repeats are stable even after growth for many generations without selective pressure. These results show promise for industrial production of heterologous gene products by yeast, in that multiple integrations of tandem expression units in combination with high copy number plasmid expression systems could be used to increase yields of heterologous gene products.

VII. TRANSCRIPTION INITIATION AND TERMINATION CHARACTERISTICS OF THE EXPRESSION SYSTEM

We have previously demonstrated that transcription of the PGK gene begins 36 nucleotides before the translation initiation codon of the gene[41] (see Figure 2). Recently, using the promoter fragment herein described, Derynck et al.[51] have expressed the cDNA for human immune interferon (IFN-γ) isolated by Gray et al.[7] Interestingly, instead of one transcription start at −36 when IFN-γ is connected to the PGK promoter fragment, this system initiates transcription at −40 and at three very minor sites further upstream to the ATG. Thus it is possible that the inserted restriction sites between the promoter fragment and inserted gene (see Figure 3), or the loss of natural sequence between −9 to −1 of the promoter (see Figure 3), or the gene itself may be affecting transcription initiation. It should be noted that the IFN-γ gene is expressed at a much lower level in YEpIPT than the 1 to 2% of yeast cell protein observed for the IFN-α or hepatitis surface antigen. It remains to be seen whether

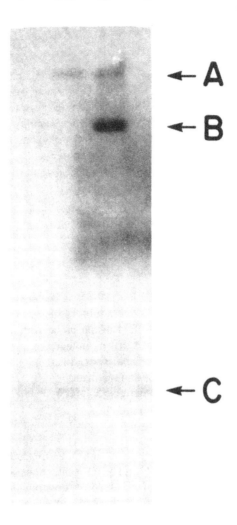

FIGURE 4.   Southern blot of genomic DNA probed with IFN-α1-*TRP1* fragment. Total genomic DNA from RH218 and three transformants (lanes 1 to 4) was digested with *Bgl*II and then electrophoresed on a 1% agarose gel. DNA was then transferred to nitrocellulose paper by the method of Southern (72) and hybridized with ^{32}P-labeled *Xba*I-*Bgl*II fragment containing sequences from the human IFN-α1 and the yeast *TRP1* genes. The lower common band results from hybridization with the indigenous *TRP1* sequences and the additional bands are due to integration events. The interferon levels as percent of total yeast protein were: 1, <0.004; 2, 0.06;, and 3, 0.2. Lane 4 contains DNA from RH218. Genomic DNA was isolated by standard procedures.[73]

multiple starts of transcription also occur for the latter genes. Recent data concerning fusions of amino-terminal coding portions of PGK structural gene to the IFN-α1 gene strongly suggest that the insertion of the *Xba*I/*Eco*RI site junction between promoter and gene has little (≤2× decrease) or no effect on expression (C.C., F.H., Hermann Oppermann, and R.H., unpublished results).[82]

Transcription termination is not a well-characterized process in yeast. As with other eukaryotes, the 3'-end of the mRNA in yeast is polyadenylated.[53] However, whether this process is signaled by the DNA or RNA sequence or is possibly an automatic process associated with mRNA chain termination has not been established. It is also possible that the polyadenylation site is not the actual termination site or release site for RNA polymerase II, but that the polymerase continues RNA synthesis through this site followed by processing or cleavage of the mRNA. Recently Zaret and Sherman[57] have compared a list of yeast gene 3'-flanking sequences and suggested two possible consensus sequences, 5'-TAG . . . TAGT . . . TTT or TAG . . . TATGT . . . TTT-3', which may have a role in transcription termination and polyadenylation; however, not all yeast genes have these consensus sequences. Bennetzen and Hall[27] have proposed a consensus sequence, 5' TAAATAA$_G^A$ 3', similar to the proposed polyadenylation signal (5' AAUAAA 3') of higher eukaryotes.[80] A comparison of many yeast 3'-flanking sequences again demonstrates great variation in this sequence. Due to the relatively unknown nature of yeast transcription termination in yeast, the following experiments will be discussed using the term as referring to the termination or processing region suggested by the overall length of the mRNA.

In our early expression system,[8] containing the *ADH1* promoter fragment—IFN-α1 cDNA—yeast *TRP1* gene, the transcription began proximal to the IFN-α1 cDNA and ended near the *TRP1* termination region.[31] The following experiments demonstrate that such a yeast 3'-flanking sequence is indeed essential for efficient expression. We constructed yeast expression systems in which the 3'-flanking DNA following the heterologous gene is either pBR322 DNA or yeast DNA. The rationale for doing this was based on the assumption that the prokaryotic DNA would probably not have proper termination, processing, or polyadenylation signals. As illustrated in Figure 5, a *Hind*III fragment containing the PGK promoter fragment, IFN-α1 gene, and the proximal region of the *TRP1* gene was inserted into plasmid YEp13[49] (plasmid B and C) or into YEp13 containing the distal region of the *TRP1* gene where yeast termination and/or processing signals occur[31] (plasmid A).

For plasmid A, transcription through the IFN-α1 gene should terminate in the distal region of the *TRP1* gene as previously described.[31] Since the plasmid B contains the other orientation of the *Hind*III fragment in YEp13, the transcript should end about 0.5 kbp from the *Hind*III site in 2 μm plasmid[50] DNA as shown. In this construction the termination and/or processing signals that normally function for the "Able" or *FLP* gene in 2 μm plasmid DNA[50] should be utilized. Plasmid C (which contains the same orientation of the *Hind*III fragment as plasmid A) lacks a 3'-flanking sequence of a yeast gene. Therefore, any transcription initiated by the PGK promoter fragment should continue through the pBR322 DNA.

The interferon expression from these three plasmids in yeast is 27 ± 10 × 10⁶ (units/ℓ of culture at an A_{660} of 1.0) for A, 14 ± 3 × 10⁶ for B, and 4 ± 2 × 10⁶ for plasmid C. Due to the large standard deviation of plasmid A activity, the difference between A and B (2-fold) may not be significant. However, the difference between A and B vs. C is 4- to 7-fold, suggesting a gross defect in plasmid C expression.

Total RNA was isolated from transformants of the yeast strain GM3C-2[19] containing each of these three plasmids. The mRNA transcript sizes were analyzed by Northern analysis,[52] as shown in Figure 6. The ³²P-labeled DNA probe which results in visualization (by hybridization) of the mRNAs was specific for IFN-α1 transcripts. Total RNA rather than polyadenylated mRNA was used for this hybridization analysis since the relative amounts of mRNA from various strains can be compared more reliably without such isolation. In Figure 6, lane A shows two different mRNA species originating from plasmid A. The majority of the mRNA is 1.5 kb in size (determined by comparison with standards) which is in agreement with the expected size of 1.45 kb. This estimate is based on the assumption that the *TRP1* gene termination region[31] was used and that polyadenylation of 50 to 100 bases[53] occurred. These data are consistent with previous results showing that such a het-

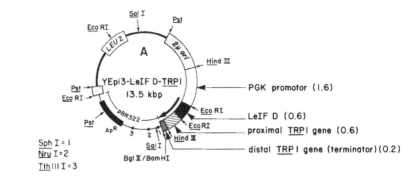

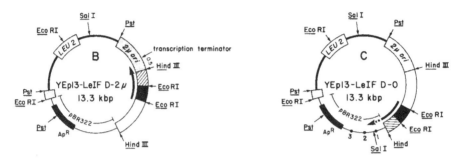

FIGURE 5.   Plasmid constructions for yeast transcription termination study. All three plasmids are based on YEp13[49] which contains the yeast *LEU2* gene and the 2 μm plasmid origin of replication. The *Hind*III fragment containing the PGK promoter fragment (*Hind*III to *Eco*RI), the IFN-α1 (or LeIF D) cDNA, and the proximal region of the *TRP1* gene was obtained from a plasmid similar to pFRS36[8] which contains the PGK promoter instead of the *ADH1* promoter. The distal portion of the *TRP1* gene had already been added to YEp13 in construction pYeHBs.[39] Plasmid constructions were done using previously described methods and placed in yeast strain GM3C-2[19] by Leu⁺ complementation using standard transformation procedures.

erologous expression system can use normal yeast transcription termination signals.[8,39] An additional mRNA species that is about 600 bases longer than the major species is found in yeast transformants containing plasmid A. It is very likely that this minor transcript terminates in the Tc^R structural gene of the pBR322 about 600 bases from the *Bgl*II/*Bam*HI junction (position 2 in Figure 5). While we have not yet determined whether this species is poly-adenylated, Hollenberg et al.[24] have shown that the mRNA for β-lactamase, made from pBR322 plasmid DNA, is polyadenylated and functional in yeast. Hence polyadenylation may use fortuitous *E. coli* DNA signals (prokaryotes do not polyadenylate mRNAs) or is a spontaneous process occurring after mRNA chain termination or chain cleavage in yeast. The inability of the *TRP1* gene 3'-flanking sequence to terminate all the hybrid mRNA may be related to the fact that only 70 bp of *TRP1* 3'-flanking sequence is present in this construction and thus, the region may be partially defective for its normal functions. However, when a similar plasmid containing the HBsAg gene instead of the IFN-α1 gene was tested, no such additional transcript was found.[39] This suggests that the gene itself affects transcription termination and/or processing.

Plasmid B produces only one transcript of 1.7 kb. This approximates the expected size of 1.75 kb as predicted for mRNA terminating in a region of 2 μm plasmid DNA at the *FLP* (or "Able") gene termination region.[50,54] We have also utilized the PGK 3'-flanking sequence for proper termination/polyadenylation of mRNAs (data not shown) and only one transcript size as with the 2 μm terminator.

As shown in lane C of Figure 6, several mRNA species were observed when transcription proceeds into pBR322 DNA. Four major transcripts of 1.7, 3.4, 6.0, and 7.0 kb were found.

plasmids producing the mRNA

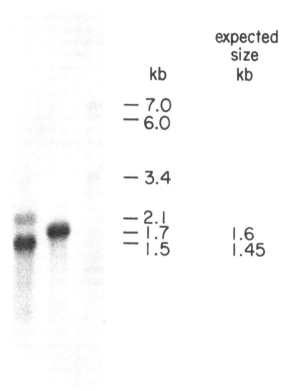

FIGURE 6. Size analysis of mRNA species produced by plasmids A, B, and C in yeast. Yeast GM3C-2[19] containing the plasmids in Figure 5 were grown in YNB-leu (39) selective medium at 30°C to an A_{660} of 1.0. Total RNA was isolated using standard procedures[74] and 50 µg of RNA was then denatured and electrophoresed in each lane of 1% agarose gel containing 6% formamide and 1 × MOPS buffer (20 mM MOPS, 5 mM NaOAc, 1 mM EDTA at pH 7.0).[75] The gel was stained with ethidium bromide using yeast ribosomal RNA[76] and HindIII cut phage γ DNA[77] as size standards. The DNA in the gel was then transferred to nitrocellulose and hybridized (Northern procedure[52]) with the 0.56 kbp EcoRI fragment containing the IFN-α1 gene[5,8] labeled with ^{32}P using the calf thymus primer method.[78]

The first transcript (1.7) ends near the SphI site (position 1 in plasmid C) in the Tc[R] structural gene about 500 nucleotides from the HindIII site. This region may also be used by plasmid A to some extent but definitely in minor amounts as compared to the other two regions. However, the region used in pBR322 by plasmid A near the NruI site (position 2) is not used in plasmid C. The NruI and SphI sites are about 400 bp apart and the next termination region (position 3) of plasmid C is 1.7 kb past the SphI site or 1.3 kb past the NruI site. This unexpected finding suggests that the termination or processing that occurs near the NruI site in plasmid A must be influenced by the yeast TRP1 3'-flanking gene signals.

Determination of the actual polyA addition sites by S1 mapping[55,56] of these mRNAs in plasmids A and C would be of great interest and would possibly give further insight into this difference.

The major transcripts for plasmid C are 6.0 and 7.0 kb and show that RNA polymerase II moves along the pBR322 DNA sequence to two stop or cleavage signals in yeast DNA flanking the *LEU2* gene from chromosome III of yeast.[14] This type of run-on transcription is reminiscent of that seen by Zaret and Sherman[57] for the yeast *cyc1-512* mutant which has a deletion of the 3'-flanking region encompassing termination or processing signals for the *CYC1* gene. They saw transcripts of 630, 850, 1350, 1450, 1650, and 2400 nucleotides for *cyc1-512* instead of only one transcript of 630 nucleotides as observed for the wild-type *CYC1* gene. The reasons for several apparent transcription stops or processing sites in their homologous and our heterologous system are still unclear. Although we have not demonstrated whether our run-on transcripts are polyadenylated, Zaret and Sherman[57] have demonstrated that all transcripts are polyadenylated and suggest that polyadenylation and transcription termination and/or processing may be coupled or that polyadenylation may be automatic at 3'-ends of mRNAs.

Finally, the relative steady-state amounts of mRNAs (see Figure 6) produced by plasmids A, B, and C are consistent with the expression levels of interferon produced by the three plasmids. As indicated earlier, yeast transformants containing plasmid A or B produce 4- to 7-fold more interferon than the transformants containing plasmid C. This difference is reflected by the relative mRNA levels produced from the three plasmids. Zaret and Sherman[57] have previously observed this reduction in expression with the deletion of a termination or processing region. They suggest that the large untranslated regions of the run-on transcripts may lead to reduced mRNA stability. They also note that the longest transcripts are present at relatively higher levels than the shorter transcripts, which seems inconsistent with this idea. We have two types (yeast or *E. coli*) of 3'-flanking DNA for our plasmid C system. In both types of DNA our data are consistent with theirs because there is more 3.4 kb transcript than the 1.7 kb transcript (ending in *E. coli* DNA) and more 7.0 kb transcript than the 6.0 kb transcript (ending in yeast DNA). This might best be explained by smaller transcripts resulting from a single large transcription unit which is unstable. If such a processing does occur then the transcript of plasmid A may just be 2.1 kb and may not be terminating in yeast DNA. A possible model for such a system would be that an efficient yeast termination region contains an RNase cleavage site as well as an RNA polymerase II release site in close proximity. In plasmid C, pBR322 encoded RNA might contain only fortuitous cleavage sites, while the cleavage sites in yeast RNA may be preferred.

VIII. YEAST TRANSCRIPTION TERMINATION OR PROCESSING SIGNALS IN MAMMALIAN cDNA SEQUENCES

Recently, Derynck et al.[51] have demonstrated a single transcript in yeast using the PGK promoter fragment system to express the cDNA[7] for IFN-γ (human immune interferon). They have shown by S1 analysis that the RNA polyadenylation site occurs at a specific location within the mammalian cDNA 182 to 184 nucleotides after the translation stop. This terminator or processing signal on the human cDNA functions as well as the yeast PGK termination region. Furthermore, they demonstrate that this region contains consensus sequences similar to that proposed by Zaret and Sherman[57] for transcription termination, processing, and polyadenylation in yeast.

We have also found such a termination or processing signal in the cDNA of the IFN-α2 (approximately 200 bp after the translation stop, data not shown). Such a phenomenon might therefore be a general occurrence. An investigation of other higher eukaryotic genes may demonstrate the evolutionary retention of this region in higher eukaryotes. Recently Tuite et al.[58] found relatively high levels of a PGK-IFN-α2 fusion protein (1 to 2% of cell protein) using an expression system without a yeast 3'-flanking sequence after the IFN-α2 cDNA. Since the 3'-flanking sequence was pBR322 DNA, one would have expected low expression if this cDNA termination or processing signal was not present, in view of our results already discussed.

IX. FUNCTION OF SPECIFIC MAMMALIAN SECRETION SIGNAL SEQUENCES IN YEAST

Using the heterologous expression system shown in Figure 3, we have expressed interferon cDNAs for either mature or pre-interferons.[59] As shown at the bottom of Figure 1, construction A contains the gene with the signal coding sequence replaced by an ATG; however, construction B contains the entire cDNA with the natural ATG preceding the coding sequence for the secretion signal peptide. Three different interferons were examined for the effect of pre-sequence vs. the effect of no pre-sequence.[59] The interferon cDNAs for IFN-α1,[5] IFN-α2,[5] and IFN-γ[7] have been previously isolated and modified for direct expression as construction A types. All of the three interferons expressed from construction A (see Figure 1) remained inside the cell; while these same interferons from construction type B (secretion signal present) resulted in interferon production in the media and the periplasmic space as well as inside the cell. Thus human secretion signals are being utilized by the lower eukaryote yeast and are resulting in secretion of protein from the yeast cell.

Since secretion from human cells involves cleavage of the signal peptide by a signal peptidase[60] during the secretion process,[61] the authors[59] examined the nature of the amino-terminal processing of two different secreted interferons from yeast. These two interferons are shown as lines 2 and 3 of Figure 7. Origins of the secretion signals are indicated on the left side while the right side shows the site recognized by the signal peptidase to give the mature secreted proteins. The pre D/A signal was made by making a hybrid between signals pre A and pre D by means of the DdeI restriction site in the DNA as designated. The interferon of line 3 (pre D/A IFN-α2) was produced as two forms in yeast media. Most of it (64%) was processed as +1 form (see Figure 7) or as processed by human cells; however, another form (−3) containing 3 amino acids of pre-sequence was also present (36%). Since pre D/A IFN-α2 is a hybrid, natural pre D IFN-α1 was also examined and again gave both the +1 form (45%) and the −3 form (47%) in yeast medium. Thus, although yeast is able to recognize the human signal sequence, the fidelity of the system is not perfect. Furthermore, the secretion is limited to about 30% of the interferon made with the rest remaining inside the cell, even though most of this material is also processed as +1 and −3 forms.

Recently Wieland et al.[62] have also examined the processing of pre-human growth hormone (pre-HGH[63]) by yeast as shown at the bottom of Figure 7. Unlike the interferon results, all of the HGH in the yeast media was properly processed +1 mature human growth hormone. However, the HGH protein remaining within the cell was not processed and retained the pre-sequence. Thus this system shows fidelity in processing of the secreted form but a defect in entering the secretion pathway.[64] In a similar fashion, Bektesh et al.[65] have expressed the mammalian secretory protein, α1-antitrypsin in yeast using the ADH1 promoter fragment. With the pre-sequence present they were unable to detect the protein product in the media or in the cell even though a transcript was made. This is reminiscent of preRGH (pre-rat growth hormone) results already discussed. Perhaps the pre-sequence leads to instability of the protein product. When they removed the α1-antitrypsin signal and replaced it with an ATG (construction B changed to A, Figure 1), the mature α1-antitrypsin protein was detected at relatively high levels. The rat growth hormone signal and the α1-antitrypsin signal may represent a third type of mammalian signal sequence which is not properly recognized by the yeast secretory apparatus and therefore leads to·degradation of the protein.

Human Cell Secretion Products

	−26	−25	−24	−23	−22	−21	−20	−19	−18	−17	−16	−15	−14	−13	−12	−11	−10	−9	−8	−7	−6	−5	−4	−3	−2	−1	+1	+2	
																DdeI site in DNA										Cleavage site			
preA				Met	Ala	Leu	Thr	Phe	Ala	Leu	Leu	Val	Ala	Leu	Leu	Val	Leu	Ser	Cys	Lys	Ser	Ser	Cys	Ser	Val	Gly	Cys	Asp	IFN-α2 continued
preD				Met	Ala	Ser	Pro	Phe	Ala	Leu	Leu	Met	Ala	Leu	Val	Val	Leu	Ser	Cys	Lys	Ser	Ser	Cys	Ser	Leu	Gly	Cys	Asp	IFN-α1 continued
preD/A				Met	Ala	Ser	Pro	Phe	Ala	Leu	Leu	Met	Ala	Leu	Val	Val	Leu	Ser	Cys	Lys	Ser	Ser	Cys	Ser	Val	Gly	Cys	Asp	IFN-α2 continued
preγ							Met	Lys	Tyr	Thr	Ser	Tyr	Ile	Leu	Ala	Phe	Gln	Leu	Cys	Ile	Val	Leu	Gly	Ser	Leu	Gly	Cys	Tyr		IFN-γ continued
preHGH	Met	Ala	Thr	Gly	Ser	Arg	Thr	Ser	Leu	Leu	Leu	Ala	Phe	Gly	Leu	Leu	Cys	Leu	Pro	Trp	Leu	Gln	Glu	Gly	Ser	Ala	Phe	Pro	HGH continued

FIGURE 7.   Amino acid sequences of human secretion signals. A comparison of the amino acid sequence of the signal prepeptide regions of human IFN-α1 (pre D), IFN-α2 (pre A), IFN-α1 (pre D/A), IFN-γ (pre γ), and HGH (pre HGH) is shown. The amino acids underlined represent differences between the amino acid sequences of pre A and pre D. The *DdeI* site indicates the junction of the D and A pre-sequences in preparation of the hybrid pre D/A pre-sequence. The cleavage site that is used in human cells during the secretion process is shown as well as the human cell secretion products (also known as the mature forms).

X. IMPLICATIONS FOR BIOTECHNOLOGY

There are several possible advantages that yeast heterologous expression systems may have over *E. coli* systems. The major advantages are the finely developed fermentation science, the possibility of glycosylation, the absence of inducible or contaminating viruses which might result in cell lysis during production, and the lack of endotoxins. So far the possibility for glycosylation looks unpromising, especially after HBsAg, which has sequences for polysaccharide addition,[66] does not appear to be glycosylated in yeast.[36,39] Since the polysaccharide additions in yeast[67] are not exactly like that for mammalian cells, such additions could be more of a disadvantage than an advantage.

A slight disadvantage of yeast, as compared to *E. coli* or other bacterial host systems, is the difficulty of breaking the cells to obtain the product. However, this difficulty may in fact be the greatest advantage of yeast in light of the secretion results herein discussed. Yeast is a sturdy microorganism capable of withstanding high hydrostatic pressure and does not easily lyse after death. There is very little cell lysis in stationary cultures and cultures grown to extremely high densities. Furthermore, the normal media proteins represent only 0.5% of the total cellular protein and consist of 5 to 8 proteins of greater than 50,000 daltons, which make up about 90% of the protein content of the media. Hence if 5% of the cellular protein would be secreted as a desirable protein product, the product would be of 90% purity. The relatively protein-free medium, combined with the resistance of yeast to external stresses, may make it an ideal system for secretion. This may be done using some natural heterologous secretion signals, with homologous yeast signal sequences[68-71,79] attached to heterologous genes, or perhaps with hybrid signals. With regard to this, we have demonstrated that homologous protein signal sequences, such as those for yeast invertase[70] and yeast α-factor,[69,79] attached to heterologous proteins result in secretion of the heterologous proteins into the culture media and proper processing.[83,84] Such systems not only allow easier purification but also produce natural products that do not begin with an amino-terminal methionine.

Finally, the future of yeast in the biotechnology industry will require yield improvements. The greatest opportunity to obtain this goal will be in better understanding the heterologous expression system shown in Figure 1 of this report. Thus far this system expresses heterologous gene products less efficiently than homologous genes. For example, the PGK gene on high copy number yeast plasmids expresses 20 to 35% of total yeast protein as PGK; however, using the same plasmids, human IFN-α cDNAs are expressed by PGK flanking sequences as only 1 to 2% of total yeast protein. Recent data demonstrate that this difference is due to lower steady state levels of mRNA associated with the IFN-α heterologous expression systems (Chen, C., and Hitzeman, R., unpublished results[51]). Whether this is due to a decreased rate of synthesis, an increased rate of degradation, or a combination of the two possibilities, is presently under investigation. Furthermore, the factors which cause this transcriptional or post-transcriptional effect must be identified in order to make yeast an industrially suitable organism for expression and secretion of heterologous gene products.

REFERENCES

1. **Goeddel, D. V., Kleid, D., Bolivar, F., Heyneker, H. L., Yansura, D. G., Crea, R., Hirose, T., Kraszewski, A., Itakura, K., and Riggs, A.,** Expression in *Escherichia coli* of chemically synthesized genes for human insulin, *Proc. Natl. Acad. Sci. U.S.A., 76,* 106, 1979.
2. **Goeddel, D. V., Heyneker, H. L., Hozumi, T., Arentzen, R., Itakura, K., Yansura, D. G., Ross, M. J., Miozzari, G., Crea, R., and Seeburg, P. H.,** Direct expression in *Escherichia coli* of a DNA sequence coding for human growth hormone, *Nature,* 281, 544, 1979.
3. **Nagata, S., Taira, H., Hall, A., Johnsrud, L., Streuli, M., Ecsodi, J., Boll, W., Cantell, K., and Weissmann, C.,** Synthesis in *Escherichia coli* of a polypeptide with human leukocyte interferon activity, *Nature,* 284, 316, 1980.
4. **Goeddel, D. V., Yelverton, E., Ullrich, A., Heyneker, H. L., Miozzari, G., Holmes, W., Seeburg, P. H., Dull, T., May, L., Stebbing, N., Crea, R., Maeda, S., McCandliss, R., Sloma, A., Tabor, J. M., Gross, M., Familletti, P. C., and Pestka, S.,** Human leukocyte interferon produced by *Escherichia coli* is biologically active, *Nature,* 287, 411, 1980.
5. **Goeddel, D. V., Leung, D. W., Dull, T. J., Gross, M., McCandliss, R., Lawn, R. M., Seeburg, P. H., Ullrich, A., Yelverton, E., and Gray, P. W.,** The structure of eight distinctive cloned human leukocyte interferon cDNAs, *Nature,* 290, 20, 1981.
6. **Goeddel, D. V., Shepard, H. M., Yelverton, E., Leung, D., Crea, R., Sloma, A., and Pestka, S.,** Synthesis of human fibroblast interferon by *Escherichia coli, Nucleic Acids Res.,* 8, 4057, 1980.
7. **Gray, P. W., Leung, D. W., Pennica, D., Yelverton, E., Najarian, R., Simonsen, C. C., Derynck, R., Sherwood, P. J., Wallace, D. M., Berger, S. L., Levinson, A. D., and Goeddel, D. V.,** Expression of human immune interferon cDNA in *Escherichia coli* and monkey cells, *Nature,* 295, 503, 1982.
8. **Hitzeman, R. A., Hagie, F. E., Levine, H. L., Goeddel, D. V., Ammerer, G., and Hall, B. D.,** Expression of a human gene for interferon in yeast, *Nature,* 293, 717, 1981.
9. **Hamer, D. H. and Leder, P.,** Expression of the chromosomal mouse B^{maj}-globin gene cloned in SV40, *Nature,* 281, 35, 1979.
10. **Palva, I., Sarvas, M., Lehtoväära, Sibakov, M., and Kääriäinen, L.,** Secretion of *Escherichia coli* beta-lactamase from *Bacilius subtilis* by the aid of alpha-amylase signal sequence, *Proc. Natl. Acad. Sci. U.S.A.,* 79, 5582, 1982.
11. **Gray, G., McKeown, K., Jones, A., Seeburg, P., and Heyneker, H.,** *Biotechnol,* 2, 161, 1984.
12. **Thompson, C. J., Ward, J. M., and Hopwood, D. A.,** Cloning of antibiotic resistance nutritional genes in streptomycetes, *J. Bacteriol.,* 151, 668, 1982.
13. **Hinnen, A., Hicks, J. B., and Fink, G. R.,** Transformation of yeast, *Proc. Natl. Acad. Sci. U.S.A.,* 75, 1929, 1978.
14. **Ratzkin, B. and Carbon, J.,** Functional expression of cloned yeast DNA in *Escherichia coli, Proc. Natl. Acad. Sci. U.S.A.,* 74, 487, 1977.
15. **Beggs, J. D.,** Transformation of a yeast by a replicating hybrid plasmid, *Nature,* 275, 104, 1978.
16. **Struhl, K., Stinchcomb, D. T., Scherer, S., and Davis, R. W.,** High frequency transformation of yeast: autonomous replication of hybrid DNA molecules, *Proc. Natl. Acad. Sci. U.S.A.,* 76, 1035, 1979.
17. **Hsiao, C. L. and Carbon, J.,** High-frequency transformation of yeast by plasmids containing the cloned yeast ARG4 gene, *Proc. Natl. Acad. Sci. U.S.A.,* 76, 3829, 1979.
18. **Kielland-Brand, M. C., Nilsson-Tillgren, T., Holmberg, S., Petersen, J. G. L., and Svenningsen, B. A.,** Transformation of yeast without the use of foreign DNA, *Carlsberg Res. Commun.,* 44, 77, 1979.
19. **Faye, G., Leung, D. W., Tatchell, K., Hall, B. D., and Smith, M.,** Deletion mapping of sequences essential for *in vivo* transcription of the iso-1-cytochrome c gene, *Proc. Natl. Acad. Sci. U.S.A.,* 78, 2258, 1981.
20. **Williamson, V. M., Bennetzen, J., Young, E. T., Nasmyth, K., and Hall, B. D.,** Isolation of the structural gene for alcohol dehydrogenase by genetic complementation in yeast, *Nature,* 283, 214, 1980.
21. **Williamson, V. M., Young, E. T., and Ciriacy, M.,** Transposable elements associated with constitutive expression of yeast alcohol dehydrogenase II., *Cell,* 23, 605, 1981.
22. **Hollenberg, C. P.,** The expression in *Saccharomyces cerevisiae* of bacterial beta-lactamase and other antibiotic resistance genes integrated in a 2-micron DNA vector, in *Extrachromosomal DNA,* Cummings, D. J., Borst, P., Dawid, I. B., Weissmann, S. M., and Fox, C. F., Eds., New York, ICN-UCLA, Symp. 15, Academic Press, 1979, 325.
23. **Chevallier, M. R. and Aigle, M.,** Qualitative detection of penicillinase produced by yeast strains carrying chimeric yeast-coli plasmids, *FEBS Lett.,* 108, 179, 1979.
24. **Hollenberg, C. P., Kustermann-Kuhn, B., Mackedonski, V., and Erhart, E.,** The expression of bacterial antibiotic resistant gene in the yeast *Saccharomyces cerevisiae,* in *Molecular Genetics in Yeast,* Wettstein, D., Friis, J., Kielland-Brandt, M., and Sterderup, A., Eds., Alfred Benzon Symposium 16, Munksgäärd, Copenhagen, 1979, 325.

25. **Cohen, J. D., Eccleshall, T. R., Needleman, R. B., Federoff, H., Buchferer, B., and Marmur, J.,** Functional expression in yeast of the *Escherichia coli* plasmid gene coding for chloramphenicol acetyltransferase, *Proc. Natl. Acad. Sci. U.S.A.,* 77, 1078, 1980.

26. **Beggs, J. D., Berg, J. V. D., Ooyen, A. V., and Weissmann, C.,** Abnormal expression of chromosomal rabbit beta-globin gene in *Saccharomyces cerevisiae, Nature,* 283, 835, 1980.

27. **Fraser, T. H. and Bruce, B. J.,** Synthesis of chicken ovalbumin in *Saccharomyces cerevisiae,* in *Microbiology — 1981,* Schlessinger, D., Ed., American Society for Microbiology, Washington, D. C., 1981, 392.

28. **Henikoff, S., Tatchell, K., Hall, B. D., and Nasmyth, K. A.,** Isolation of a gene from *Drosophilia* by complementation in yeast, *Nature,* 289, 33, 1981.

29. **Bennetzen, J. L. and Hall, B. D.,** The primary structure of the *Saccharomyces cerevisiae* gene for alcohol dehydrogenase, *J. Biol. Chem.,* 257, 3018, 1982.

30. **Stinchcombe, D. T., Struhl, K., and Davis, R. W.,** Isolation and characterization of a yeast chromosomal replicator, *Nature,* 282, 39, 1979.

31. **Tschumper, G. and Carbon, J.,** Sequence of a yeast DNA fragment containing a chromosomal replicator and the *Trp1* gene, *Gene,* 10, 157, 1980.

32. **Stewart, W. E., II,** Interferon assays, in *The Interferon System,* Springer, New York, 1979, 13.

33. **Carbon, J., Ratzkin, B., Clarke, L., and Richardson, D.,** The expression of cloned eukaryotic DNA in prokaryotes, *Brookhaven Symp. Biol.,* 29, 277, 1977.

34. **Ammerer, G., Hitzeman, R., Hagie, F., Barta, A., and Hall, B. D.,** The functional expression of mammalian genes in yeast, in *Recombinant DNA,* Proceedings of the 3rd Cleveland Symp. on Macromolecules, Walton, A. G., Ed., Elsevier Scientific Publishing Co., Amsterdam, Netherlands, 1981, 185.

35. **Bennetzen, J. L. and Hall, B. D.,** Codon selection in yeast, *J. Biol. Chem.,* 257, 3026, 1982.

36. **Valenzuela, P., Medina, A., Rütter, W. J., Ammerer, G., and Hall, B. D.,** Synthesis and assembly of hepatitis B virus surface antigen particles in yeast, *Nature,* 298, 347, 1982.

37. **Miyanohara, A., Toh-e, A., Nozaki, C., Hamada, F., Ohtomo, N., and Matsubara, K.,** Expression of hepatitis B surface antigen gene in yeast, *Proc. Natl. Acad. Sci. U.S.A.,* 80, 1, 1983.

38. **Kramer, R. A. and Andersen, N.,** Isolation of yeast genes with mRNA levels controlled by phosphate concentration, *Proc. Natl. Acad. Sci. U.S.A.,* 77, 6541, 1980.

39. **Hitzeman, R. A., Chen, C. Y., Hagie, F. E., Patzer, E. J., Liu, C., Estell, D. A., Miller, J. V., Yaffe, A., Kleid, D. G., Levinson, A. D., and Oppermann, H.,** *Nucleic Acids Res.,* 11, 2745, 1983.

40. **Hitzeman, R. A., Clarke, L., and Carbon, J.,** Isolation and characterization of the yeast 3-phosphoglycerokinase gene (PGK) by an immunological screening technique, *J. Biol. Chem.,* 255, 12073, 1980.

41. **Hitzeman, R. A., Hagie, F. E., Hayflick, J. S., Chen, C. Y., Seeburg, P. H., and Derynck, R.,** The primary structure of *Sacharomyces cerevisiae* gene for 3-phosphoglycerate kinase, *Nucleic Acids Res.,* 10, 7791, 1982.

42. **Miozzari, G. F., Niederberger, P., and Hütter, R.,** Tryptophan biosynthesis in *Saccharomyces cerevisiae*: control of the flux through the pathway, *J. Bacteriol.,* 134, 48, 1978.

43. **Huang, I. Y., Welch, C. D., and Yoshida, A.,** Complete amino acid sequence of human phosphoglycerate kinase. Cyanogen bromide peptides and complete amino acid sequence, *J. Biol. Chem.,* 255, 6412, 1980.

44. **Banks, R. D., Blake, C. C. F., Evans, P. R., Haser, R., Rice, D. W., Hardy, G. W., Merrett, M., and Phillips, A. W.,** Sequence, structure and activity of phosphoglycerate kinase: a possible hinge-bending enzyme, *Nature,* 279, 773, 1979.

45. **Guarente, L., Yocum, R. R., and Gifford, P.,** *GAL10-CYC1* hybrid yeast promoter identifies the *GAL4* regulatory region as an upstream site, *Proc. Natl. Acad. Sci. U.S.A.,* 79, 7410, 1982.

46. **Struhl, K.,** The yeast *his3* promoter contains at least two distinct elements, *Proc. Natl. Acad. Sci. U.S.A.,* 79, 7385, 1982.

47. **Beier, D. R., and Young, E. T.,** Characterization of a regulatory region upstream of the *ADR2* locus of *Saccharomyces cerevisiae, Nature,* 300, 724, 1982.

48. **Bolivar, F., Rodriguez, R. L., Greene, P. J., Betlach, M. C., Heyneker, H. L., and Boyer, H. W.,** Construction and characterization of new cloning vehicles. II. A multipurpose cloning system, *Gene,* 2, 95, 1977.

49. **Broach, J. R., Strathern, J. N., and Hicks, J. B.,** Transformation in yeast: development of a hybrid cloning vector and isolation of the *CAN1* gene, *Gene,* 8, 121, 1979.

50. **Hartley, J. L. and Donelson, J. E.,** Nucleotide sequence of the yeast plasmid, *Nature,* 286, 860, 1980.

51. **Derynck, R., Singh, A., and Goeddel, D. V.,** Expression of the human interferon-gamma cDNA in yeast, *Nucleic Acids Res.,* 11, 1819, 1983.

52. **Thomas, P. S.,** Hybridization of denatured RNA and small DNA fragments transferred to nitrocellulose, *Proc. Natl. Acad. Sci. U.S.A.,* 77, 5201, 1980.

53. **McLaughlin, C. S., Warner, J. R., Edmonds, M., Nakazato, H., and Vaughan, M. H.,** Polyadenylic acid sequences in yeast messenger ribonucleic acid, *J. Biol. Chem.,* 248, 1466, 1973.

54. **Broach, J. R. and Hicks, J. B.**, Replication and recombination functions associated with the yeast plasmid, 2 mu circle, *Cell*, 21, 501, 1980.
55. **Berk, A. J. and Sharp, P. A.**, Spliced early mRNAs of simian virus 40, *Proc. Natl. Acad. Sci. U.S.A.*, 75, 1274, 1978.
56. **Mantei, N., Schwarzstein, M., Streuli, M., Panem, S., Nagata, S., and Weissmann, C.**, The nucleotide sequence of a cloned human leukocyte interferon cDNA, *Gene*, 10, 1, 1980.
57. **Zaret, K. S. and Sherman, F.**, DNA sequences required for efficient transcription termination in yeast, *Cell*, 28, 563, 1982.
58. **Tuite, M. F., Dobson, M. J., Roberts, N. A., King, R. M., Burke, D. C., Kingsman, S. M., and Kingsman, A. J.**, Regulated high-efficiency expression of human interferon-alpha in *Saccharomyces cerevisiae*, *EMBO*, 1, 603, 1982.
59. **Hitzeman, R. A., Leung, D. W., Perry, L. J., Kohr, W. J., Levine, H. L., and Goeddel, D. V.**, Secretion of human interferon by yeast, *Science*, 219, 620, 1983.
60. **Jackson, R. C. and Blobel, G.**, Post-translational cleavage of presecretory proteins with an extract of rough microsomes from dog pancreas containing signal peptidase activity, *Proc. Natl. Acad. Sci. U.S.A.*, 74, 5598, 1977.
61. **Palade, G.**, Intracellular aspects of the process of protein synthesis, *Science*, 189, 347, 1975.
62. **Wieland, A., Miller, J., Estell, D., Jones, A., Heyneker, H., and Hitzeman, R.**, unpublished results.
63. **Goodman, H. M. et al.**, Structure and expression in bacteria of growth hormone genes, in *Specific Eukaryotic Genes*, Engberg, J., Klenow, H., and Leick, V., Eds., Munskagäärd, Copenhagen, 1979, 179.
64. **Schekman, R. and Novick, P.**, The secretory process and yeast cell-surface assembly, In *The Molecular Biology of the Yeast Saccharomyces*, Vol II, Strathern, J. N., Jones, E. W., and Broach, J. R., Eds., Cold Spring Harbor Laboratory, Cold Spring Harbor, New York, 1982, 361.
65. **Bektesh, S., Kurachi, K., Long, G., and Hall, B.**, manuscript in preparation.
66. **Valenzuela, P., Gray, P., Quiroga, M., Zaldivar, J., Goodman, H. M., and Rütter, W. J.**, Nucleotide sequence of the gene coding for the major protein of hepatitis B virus surface antigen, *Nature*, 280, 815, 1979.
67. **Ballou, C. E.**, Yeast cell wall and cell surface, in *The Molecular Biology of the Yeast Saccharomyces*, Vol II Strathern, J. N., Jones, E. W., and Broach, J. R., Eds., Cold Spring Harbor Laboratory, Cold Spring Harbor, New York, 1982, 335.
68. **Meyhack, B., Bajwa, W., Rudolph, H., and Hinnen, A.**, Two yeast acid phosphatase structural genes are the result of a tandem duplication and show different degrees of homology in their promoter and coding sequences, *EMBO*, 1, 675, 1982.
69. **Kurjan, J. and Herskowitz, I.**, Structure of a yeast pheromone gene (MF alpha): a putative alpha-factor precursor contains four tandem copies of a mature alpha-factor, *Cell*, 30, 933, 1982.
70. **Carlson, M. and Botstein, D.**, Two differentially regulated mRNAs with different 5' ends in code secreted with intracellular forms of yeast invertase, *Cell*, 28, 145, 1982.
71. **Fogel, S. and Welsh, J. W.**, Tandem gene amplification mediates copper resistance in yeast, *Proc. Natl. Acad. Sci. U.S.A.*, 79, 5342, 1982.
72. **Southern, E. M.**, Detection of specific sequences among DNA fragments separated by gel electrophoresis, *J. Mol. Biol.*, 98, 503, 1975.
73. **Smith, D. and Halvorson, H. O.**, The isolation of DNA for yeast, *Methods Enzymol.*, 12, 538, 1967.
74. **Zitomer, R. S. and Hall, B. D.**, Yeast cytochrome c messenger RNA. *In vitro* translation and immunoprecipitation of the *CYC1* gene product, *J. Biol. Chem.*, 251, 6320, 1976.
75. **Dobner, P. R., Kawasaki, E. S., Yu, L. U., and Bancroft, F. C.**, Thyroid or glucocorticoid hormone induces pre-growth-hormone mRNA and its probable nuclear precursor in rat pituitary cells, *Proc. Natl. Acad. Sci. U.S.A.*, 78, 2230, 1981.
76. **Warner, J. R.**, The yeast ribosome: structure, function, and synthesis, in *The Molecular Biology of the Yeast Saccharomyces*, Vol II Strathern, J. N., Jones, E. W., and Broach, J. R., Eds., Cold Spring Harbor Laboratory, New York, 1982, 529.
77. **Szybalski, E. H. and Szybalski, W.**, A comprehensive molecular map of bacteriophage lambda, *Gene*, 7, 217, 1979.
78. **Taylor, J. M., Illmensee, R., and Summers, J.**, Efficient transcription of RNA into DNA by avian sarcoma virus polymerase, *Biochem. Biophys. Acta*, 442, 324, 1976.
79. **Singh, A., Chen, E. Y., Lygovoy, J. M., Chang, C. N., Hitzeman, R. A., and Seeburg, P. H.**, *Saccharomyces cerevisiae* contain two discrete genes coding for the alpha-factor pheromone, *Nucleic Acids Res.*, 12, 4049, 1983.
80. **Proudfoot, N. J., Cheng, C. C., and Brownlee, G. G.**, Sequence analysis of eukaryotic mRNA, *Progr. Nucleic Acid Res. Mol. Biol.*, 19, 123, 1976.
81. **Hall, B.**, personal communication.
82. **Chen, C., Hagie, F., Oppermann, H., and Hitzeman, R.**, unpublished results.

83. **Singh, A., Chang, C. N., Lugovoy, J. M., Matteucci, M. D., and Hitzeman, R. A.,** Strategies for the production of pharmaceutical proteins in microorganisms, *Proc. XV Intl. Congr. Genet.,* in press.
84. **Singh, A., Lugovoy, J. M., Kohr, W. J., and Perry, J. L.,** Synthesis and Secretion of mammalian proteins in yeast using α-factor promotor and presequences, *Proc. XV Intl. Congr. Genet.,* in press.

Chapter 4

BACKGROUND TO HUMAN INTERFERON

Norwood O. Hill

I. RELEVANCE TO THE BIOTECHNOLOGY REVOLUTION

Interferons have special significance to Recombinant DNA technology as prototype modifers of immune response. In addition, interest in therapeutic potential of interferon against cancer and virus diseases has given impetus to the emerging Recombinant DNA industry. The interferon gene was one of the earliest isolated and expressed for production of its gene product with a goal of therapeutic trials.[1,2] As a result of Recombinant DNA methods, we have learned that there are multiple alpha interferon genes.[2] In addition, there are probably at least two genes for the beta interferons,[3] and possibly no more than one for gamma interferon.[4]

The necessity for so many different interferons is not yet understood. But, if we encounter such complexity with other active products of the immune system, the field of immunologically active proteins will be one of the most fruitful for study.

It is probable that many of these other immune system response modifers will also be targets for gene cloning and then production by Recombinant DNA methods. A therapeutic goal to manipulate the immune system seems certain to underly many of these efforts as well.

Other biotechnology developments, such as monoclonal antibodies, will also play a significant role in targeting Recombinant DNA efforts. Monoclonal antibodies are enabling immunologists to identify subpopulations of leukocytes.[5] Antibodies against specific proteins of interest can identify the cells which produce selected biological response modifiers.[6] Moreover, receptor antibodies can enable studies of target cell interaction, as well as enumeration of receptors on different types of target cells.[7] Such techniques should result in a more complete understanding of immune mechanisms. More importantly, the outcome of these studies of immune mechanisms will determine some of the priorities for cloning of the genes for peptide immune mediators.

In addition to the production of interesting proteins in microorganisms, Recombinant DNA technology may enable a greater understanding of genetic factors involved in immune mediated diseases. The knowledge of immunopathology derived and the immunopharmaceutical use of human proteins produced by Recombinant DNA technology seems quite likely to be significant in the therapy of many human diseases.

II. DEVELOPMENT OF INTERFERON AND EARLY CLINICAL TRIALS

Isaacs and Lindenmann discovered interferon in 1956 while pursuing an interest in the phenomenon of viral interference. Cells infected with one virus were able to resist superinfection with a second virus. The extracellular material obtained from cultures infected by the first virus was termed "the interferon" because it could transmit viral resistance to uninfected cells.[8] The possibility that interferon, a distinct antiviral protein, might have pharmaceutical implications was recognized early.[9] However, years of careful research by several investigators were necessary before meaningful clinical trials could begin. Gresser reported in 1961 that human leukocytes could produce interferon in response to Sendai virus.[10] Subsequently, Gresser et al. reported activity of mouse interferon against spontaneous, as well as transplanted, malignancy.[11] In 1964, Wheelock et al. attempted to use viruses as interferon inducers in a terminal acute myelogenous leukemia patient.[12] Several different types of viruses were given in sequence by the intravenous route. Although the patient died, leukemic cells could not be found in the bone marrow at autopsy.

In 1966, low titer leukocyte interferon was administered to patients with acute leukemia by Falcoff et al.[13] However, the amounts of interferon probably did not exceed a few thousand units per day. This is a dose which we now know would be too low to expect any therapeutic response.

Meaningful clinical trials awaited development of a practical method to produce sufficient quantities of interferon. This was accomplished by Cantell et al. who developed such a method after nearly a decade of painstaking research.[14] Their production of interferon in Helsinki, Finland led to the first prophylactic study of interferon in osteogenic sarcoma by Hans Strander et al. of Stockholm, Sweden.[15] Prior to returning to Stockholm, Strander spent several years collaborating with Cantell in Helsinki toward the development of the production process. Initial reports from the Strander study suggested that patients who received interferon after removal of a solitary primary bone tumor, had less than the expected instance of tumor metastases to lungs at 6 months. After primary resection of the tumor, patients received 3 million units of leukocyte interferon intramuscularly for 30 days followed by 3 million units three times a week for the next 17 months. Fourteen of these patients were treated from 1972 to 1975 and were compared to thrity-three historical controls. Of thirteen patients followed more than 6 months, none developed metastases during this 6 month period. In contrast, fifteen of thirty-three historical controls developed metastases within 6 months. The differences were considered statistically significant. Due to the design without prospective randomization, the Strander report was cautiously received. Nevertheless, the report stimulated wider interest in expanded production of interferon for more extensive clinical trials. Subsequently, reports of possible antitumor activity from other sources appeared. In 1975, Habif reported observations on intralesional interferon mediated effects in skin recurrences of breast cancer.[16] In two of four patients treated intralesionally for skin recurrences of breast cancer, lymphocyte-mediated antitumor cytotoxicity appeared to increase. There was some suggestion of individual tumor nodule regression. In 1977, Jereb et al. reported regression of malignancy with local application including injection into malignant effusions.[17] In breast cancer with pleural effusions, intrapleural injection resolved effusions in four of six patients for periods of 1 month to more than 5 months. They also suggest that regressions occurred with local application of interferon to cervical carcinoma by a pessary application.[18]

In 1977 and 1978, reports by Strander et al. and Hill et al. indicated objective regression of established malignancy in the case of multiple myeloma and leukemia respectively.[19-21] Preliminary reports by Gutterman et al. and Merigan et al. also indicated objective regressions in patients with breast cancer, myeloma, and nodular lymphoma.[22] As a consequence, the American Cancer Society decided to announce a 2 million dollar grant for interferon research, as well as to publicize the reasons for that support. This occurrence in late 1978 considerably increased interest in funding interferon research, as well as the pace of interferon research.

While interest centered on the possible anti-malignant effects of interferon, observations in patients with virus diseases were also of interest. Activity against varicella-zoster,[23] herpes simplex keratitis,[24,25] cytomegalovirus,[26] and neonatal rubella were all reported.[27] Activity was also reported against a form of chronic active hepatitis caused by the hepatitis B virus.[28] In this study, Greenberg et al. indicated that interferon therapy could reduce DNA polymerase activity, as well as result in an improvement of liver function parameters. Chronic active hepatitis caused by the hepatitis B virus is a problem of worldwide scope. Indeed, in both number of cases and severity of effects, it rivals many cancers. Taken together, the effects of interferon in cancer as well as virus diseases has further increased research interest.

III. ALPHA, BETA, AND GAMMA INTERFERONS

Attempts to manufacture interferon on a large scale were undertaken. Interest in additional types of interferons expanded. The designation of type I (virus-induced) interferon and type 2 (mitogen-induced) interferon was changed to indicate apparent differences in molecular species. Interferons are now designated as alpha (leukocyte), beta (fibroepithelial), and gamma (immune) interferons. Sophisticated chemical purifications of the interferons reveal multiple species. Indeed, Berg et al. have reported the purification of thirteen alpha interferon species to homogeneity.[29] The multiplicity of alpha interferons was also confirmed by the isolation of multiple alpha genes.[2] At least two beta genes have also been isolated by Recombinant DNA techniques.[3] So far, only one gamma interferon gene has been identified.[4] However, kinetics of interferon production as, for example, reported by Osther et al. indicate that there is a variety of mitogen-stimulated interferon which appears within the first 12 to 18 hr and a second peak of activity that occurs after 48 hr.[30] This may suggest that there is more than one gamma interferon species.

The multiplicity of interferons has considerably complicated the picture for both basic science and clinical investigators. Whereas interferon produced from leukocytes may contain the multiple species of alpha interferon as well as some other biologically active impurities, Recombinant DNA methods enable the production of just one species at a time. By comparison, it should be possible to determine whether a single species of interferon suffices for treatment or whether, in somme circumstances, a combination is more desirable. In addition, it should allow some elucidation of the separate function of different alpha interferons if they, indeed, have separate functions.

Separate exploration of the beta interferons needs to be undertaken in order to determine if they have a distinctly different spectrum of usefulness from the alpha interferons. Beta interferon is known to differ pharmacologically. Unlike alpha interferon, intramuscular administration of modest doses of beta interferon does not lead to consistently detectable serum levels.[31] This may explain why some early intramuscular studies with beta interferon did not appear to give antitumor effects. In contrast, intravenous administration of beta interferon has given evidence of tumor regression in nasopharyngeal carcinoma, laryngeal papilloma, and several other malignancies.[32-34] In addition, beta interferon has been reported to have beneficial effects in multiple sclerosis where the interferon is injected into the spinal fluid.[35]

As to gamma interferon, it is produced by a gene distinctly unrelated in sequence to alpha or beta interferon genes. Gamma interferon is of special clinical interest because it appears to have greater antitumor effects in mice than either alpha or beta interferon.[36] Moreover, gamma interferon seems to potentiate effects of alpha and beta interferons in the therapy of mouse tumors.[37]

Clinical trials with human gamma interferon produced in leukocyte cultures have recently gotten underway. From these trials have come preliminary reports of objective tumor regression with this human protein. Osther et al. reported that gamma interferon was safely tolerated in doses from 10 thousand to 1 million units every other day in a Phase I trial.[30] Moreover, they reported short term objective regressions in two of three patients with breast cancer, one of two with malignant melanoma, and stabilization for at least 5 months in one of two patients with cancer of the colon. Gutterman et al. also noted objective regressions in at least one patient with breast cancer, and one with renal carcinoma metastatic to skin when gamma interferon was given intramuscularly.[38] These findings were noted despite the fact that blood levels of gamma interferon were not significantly detectable after the intramuscular administration. This may suggest that gamma interferon stimulates secondary immunologic phenomena.

IV. INTERFERON AND IMMUNOMODULATION

With regard to alpha and beta interferons, many of their antitumor effects take place at rather nominal doses. These doses would not be expected to challenge many cellular receptors in body tissues.[39] For example, the dose of 3 million units per day would be expected to challenge on the order of three receptors per cell in the body per day assuming an even distribution throughout all tissues. This would be exceptionally small on the scale of receptor interactions. On the other hand, some clinical effects, such as those against acute lymphatic leukemia, usually require high doses that would be expected to lead to a significant receptor challenge. These studies carried out at approximately 1 to 2 million units /kg/day, would challenge on the order of 80 to 160 receptors per cell per day.

These differences suggest that different mechanisms of action may be at play in the response of different tumors. At low dose one would be more inclined to attribute antitumor effects to immunomodulating properties of interferon. These may be of great importance if interferon is able to trigger other antitumor activities of the immune system to greater activity. However, direct effects on tumor cells may be of importance and may, on occasion, require high receptor challenge, as in the case of leukemia. With regard to immunomodulation, studies of the influence of interferon at various doses in vivo and in vitro on antibody-dependent cell cytotoxicity (ADCC), natural killer cell function, macrophage activation, and of the influences on suppressor cell function are being actively pursued in a number of laboratories.[40-44] The importance of elucidating the beneficial immunologic effects of interferon and learning how to augment them may be of future importance in interferon therapy as well. Augmentation of effects between alpha or beta interferons and gamma interferon in mice is an example. The possibility that interferons will be potentiated by other immunologic proteins, such as lymphotoxin, interleuken 2, or macrophage activating factor, must eventually be explored. Possible antitumor effects of lymphotoxin have already been reported in preclinical and early clinical trials in humans by Khan et al.[45,46] Interleuken 2 (T cell growth factor) is also entering early clinical trials.[47] After the appropriate single agent studies of these substances, it seems quite likely that they may be given in combination with one another or with one of the interferons.

Recent examples of the immune potentiation of the effect of interferon against malignant melanoma are contained in the reports of Borgström et al. and Hill et al.[48,49] They have combined interferon and cimetidine in malignant melanoma, a form of malignancy previously resistent to interferon. Of a total of 31 patients so far reported from the two studies, 11 have had objective regression. This includes five complete responses, five with greater than 50% decrease in tumor size (partial response), and one with 25 to 50% decrease in tumor size (minor response). The response rate of over 30% suggests improvement over the response rates of about 20% expected with single agent chemotherapy. Cimetidine, a histamine antagonist at the H2 receptor site, is known to inhibit histamine activated suppressor cell function.[50] This leads to in vitro improvement in natural killer cell function, as well as improved lymphocyte response to mitogens.[51-54] In addition, cimetidine has been reported to improve survival of separate groups of mice bearing at least two different tumors.[55,56] Clearly, the combination of other immunomodulators with interferon, whether they be other products of white cells or synthetic drugs, is of interest.

Therapeutically, the number of diseases which can be influenced favorably by interferon constitutes an impressive list. This includes both virus illnesses and a substantial list of malignancies. Despite several years of study, however, the ultimate role of interferon is not yet clear. In malignancy, with a few exceptions, interferon alone is not yet considered to be a standard treatment for patients, expecially in earlier stages of disease. But, there are exceptions. These would include metastatic renal carcinoma, a malignancy with a poor response rate to radiation or chemotherapy. Interferon may be the best single agent for this disease. Gutterman et al., reporting on a series of 37 patients, found a response rate of 40%.[57] This included all patients having greater than 50% decrease in size of all measurable tumors. In contrast, response rates to single agent chemotherapy are generally below 20%.[58] Interferon and cimetidine, as previously mentioned, may now be an optimum treatment choice for certain forms of recurrent malignant melanoma.[48,49] In addition, the benign tumor, juvenile laryngeal papilloma, is best treated with interferon. Although relatively rare, this condition can be fatal due to obstruction of the respiratory tree. Strander and Haglund et al. have demonstrated the long-term benefit of interferon treatment for this tumor.[59] Some children in Sweden treated for more than 8 years have been successfully withdrawn from the therapy with interferon without recurrence of the papillomas.

It seems clear that the ability to pursue the clinical investigation of interferon has been strengthened through Recombinant DNA technology. Now that substantial quantities of alpha interferon are available for testing, studies of mechanisms, optimization of dose, and incidence of response of various tumors, can be carried out on a reasonable basis. Similarly, we can expect an extension of beta interferon trials and the beginning of trials with cloned gamma interferon.

The potential for combining the alpha interferons or beta interferon with gamma interferon is suggested by that beneficial combination in the mouse models. Moreover, the potential for utilizing other biological response modifiers such as interleuken 2, lymphotoxin, or macrophage activating factors, among others, may hold therapeutic promise for the future. Central to most of these studies will be the Recombinant DNA technology, both for the production of the products needed and for the elucidation of the basic genetics and molecular mechanisms involved in their actions.

REFERENCES

1. **Nagata, S., Taira, H., Hall, A., Johnsrud, L., Streuli, M., Escödi, J., Boll, W., Cantell, K., and Weissmann, C.,** Synthesis in *E. coli* of a polypeptide with human leukocyte interferon activity, *Nature*, 284, 316, 1980.
2. **Goeddel, D. V., Leung, D. W., Dull, T. J., Gross, M., Lawn, R. M., McCandliss, R., Seeburg, P. H., Ullrich, A., Yelverton, E., and Gray, P. W.,** The structure of eight distinct cloned human leukocyte interferon DNAs, *Nature*, 290, 20, 1981.
3. **Weissenbach, J., Chernajovsky, Y., Zeevi, M., Shulman, L., Soveg, H., Niv, U., Wallach, D., Perncaudet, M., Trollary, P., and Revel, M.,** Two interferon mRNA's in human fibroblasts: *in vitro* translation and *Escherechia coli* cloning studies, *Proc. Natl. Acad. Sci. U.S.A.,* 77, 7152, 1980.
4. **Gray, P. W., Leung, D. W., Pennica, D., Yelverton, S., Najarian, R., Simonsen, S. S., Derynck, R., Sherwood, P. J. Wallace, D. M., Berger, S. L., Levinson, A. D., and Goeddel, D. V.,** Expression of human immune interferon cDNA E. coli and monkey cells, *Nature*, 295, 503, 1982.
5. **Janossy, G., Thomas, J., and Pizzolo, G.,** The analysis of lymphoid subpopulations in normal and malignant tissues by immunofluorescent technique, *J. Cancer Res. Clin. Oncol.,* 101, 1, 1981.
6. **Secher, D., Burke, D., and Cantell, K.,** Binding of Anti-alpha Interferon Antibody to Monocytes from Cultures Producing Interferon, Poster, International Meeting on the Biology of the Interferon System, Rotterdam, Holland, April, 1981.
7. **Kamarck, M. E., Slate, D. L., D'Eustachio, P., Barel, B., and Ruddle, R. H.,** Genetic control of the system, *Fed. Proc. Fed. Am. Soc. Exp. Biol.,* 40, 1051.
8. **Isaacs, A., and Lindenmann, J.,** Virus interference. 1. The interferon, *Proc. R. Soc. Lond.* (Biol.), 147, 258, 1957.
9. **Isaacs, A.,** Interferon: The prospects, *The Practitioner,* 183, 601, 1959.
10. **Gresser, I.,** Induction by Sendai virus of non-transmissible cytopathic changes associated with rapid and marked production of Interferon, *Proc. Soc. Exptl. Biol. Med.,* 108, 303, 1961.
11. **Gresser, I. and Bouraili, C.,** Anti-tumor effects of interferon preparation in mice, *J. Natl. Cancer Inst.,* 45, 365, 1970.
12. **Wheelock, E. F. and Dingle, J. H.,** Observations on the repeated administration of viruses to a patient with acute leukemia, *N. Engl. J. Med.,* 271, 13, 645, 1964.
13. **Falcoff, E. et al.,** Mass production, partial purification, and characterization on interferon destined for therapeutic tests in humans, *Ann. L'Inst. Pasteur,* 562, 1966.
14. **Cantell, K., Hirvonen, S., Mogensen, K. T., and Pyhälä, L.,** Human leukocyte interferon: production, purification, stability, and animal experiments, in *in vitro*, Monograph 3, 35, 139, 1977.

15. **Strander, H., Cantell, K., Jakobsson, P. A., Nilsonn, U. and Soderberg, G.,** Exogenous interferon therapy of osteogenic sarcoma, *Oerhop. Scand.,* 45, 958, 1974.

16. **Habif, D.,** *International Workshop on Interferon in the Treatment of Cancer,* 49, 1975.

17. **Jereb, B., Cervek, J., Us-Krasovec, M., Ikic, D., and Soos, E.,** Intrapleural Application of Human Leukocyte Interferon (HLI) in Breast Cancer Patients with Unilateral Pleural Carcinomas, *Proc. Symposium on Preparation, Standardization and Clinical Use of Interferon.,* Zagreb, 1977.

18. **Krusic, J., Ikic, D., Knezevic, M., Soos, E., Rode, B., Jusic, D., and Maracic, Z.,** Clinical and Histologic Findings After Local Application of Human Leukocyte Interferon in Patients with Cervical Cancer, *Proc. Symposium on Preparation, Standardization and Clinical Use of Interferon,* Zagreb, 1977.

19. **Mellstedt, H., Björkholm, M., Johansson, B., Ahre, A., Holm, G., Strander, H.,** Interferon therapy in myelomatosis, *Lancet,* 1979, 245.

20. **Hill, N. O., Loeb, E., Pardue, A. S., Dorn, G. L., Khan, A., and Hill, J. M.,** Effects of High Dose Human Leukocyte Interferon Administration in Acute Lymphocytic Leukemia, Presented before the South Central Association of Blood Banks, Little Rock, Ark., February, 1978.

21. **Hill, N. O., Loeb, E., Pardue, A. S., Dorn, G. L., Khan, A., and Hill, J. M.,** Response of acute leukemia to leukocyte interferon, *J. Clin. Hematol. and Oncol.,* 9, 1, 137, 1979.

22. **Gutterman, J. U., Blumenschein, G. R., Alexanian, R., Yap, H. Y., Buzclar, A. U., and Peska, S.,** Leukocyte interferon-induced tumor Regression in human metastatic breast cancer, multiple myeloma, and malignant lymphoma, *Ann. Intern. Med.,* 93, 399, 1980.

23. **Merigan, T. C. et al,** Human leukocyte interferon for the treatment of herpes zoster in patients with cancer, *N. Engl. J. Med.,* 298, 18, 1978.

24. **Sundmacher, R., Neumann-Haefelin, D., and Cantell, K.,** Interferon treatment of dendritic keratitis, *Lancet,* 26, 1406, 1976.

25. **Sundmacher, R., Neumann-Haefelin, D., Cantell, K.,** Successful treatment of dendritic keratitis with human leukocyte interferon, *Albrecht v. Graefes Arch. Klin. Exp. Opthal.,* 201, 39, 1976b.

26. **Emödi, G., O'Reilly, R., Muller, A., Everson, T. T., Binswanger, U., and Just, M.,** Effect of human exogenous leukocyte interferon in cytomegolovirus infections, *J. Infect. Dis.,* 133, (A), 199, 1976.

27. **Larsson, A., Forsgren, M., Hard, S., Strander, H., and Cantell, K.,** Administration of interferon to an infant with congenital rubella syndrome involving persistent viremia and cutaneous vasculitis, *Acta Pediat. Scand.,* 65, 105, 1976.

28. **Greenberg, H. B. et al,** "Effect of human leukocyte interferon on hepatitis B virus infection in patients with chronic active hepatitis, *N. Engl. J. Med.,* 295, 10, 1976.

29. **Berg, K. and Heron, I.,** Human Leukocyte Interferon Comprises a Continuum of 13 Interferon Species, in *Human Lymphokines: Biological Response Modifiers, Khan, A., and Hill, N. O.* Eds., Academic Pres, New York, 1982, 397.

30. **Osther, K., Georgiades, J., Hilario, R., Hill, R. W., Pardue, A., Khan, A., Aleman, C., Hill, J. M., and Hill, N. O.,** Human Gamma Interferon-A Phase I Trial, Presented before the 3rd Intr. Cong. for Interferon Res., Miami, Fla., November, 1982.

31. **Billiau, A., DeSomer, P., Edy, V. G., DeClercq, E., and Heremans, H.,** Human fibroblast interferon for clinical trials: pharmacokinetics and tolerability in experimental animals and humans, *Antimicrob. Agents Chemother.,* 16, 56, 1979.

32. **Treuner, J., Niethammer, D., Dannecker, G., Hagmann, R., Neef, V., and Hofschneider, P. H.,** Successful treatment of nasopharyngeal carcinoma with interferon, *Lancet,* 1, 817, 1980.

33. **Schouten, T. J. and Bos, J. H.,** "Interferon voor de behandeling var Juveniele Papillomatosis van de Larynx", *Ned T Geneeskd,* 124, 1650, 1980.

34. **Billiau, A., DeSomer, P., Schellekens, H., and Weimar, W.,** Review: the clinical value of interferon-A clinical appraisal, *Neth. J. Med.,* 24, 72, 1981.

35. **Jacobs, L. et al.,** Intrathecal interferon reduces exacerbations of multiple sclerosis, *Science,* 214, 1026, 1981.

36. **Brysk, M. M., Tschen, E. L., Hudson, R. D., Smith, E. B., Fleischmann, W. R., Jr., and Black, H. S.,** The activity of interferon on ultraviolet light-induced squamous cell carcinomas in mice, *J. Am. Acad. Dermatol.,* (1), 61, 1981.

37. **Fleischmann, W. R., Jr., Kleyn, K. M., and Baron, S.,** Potentiation of antitumor effect of virus-induced interferon by mouse immune interferon preparations, *J. Am. Acad. Dermatol,* 65, 963, 1981.

38. **Gutterman, J. U., Jordan, G. W., Rios, A., Quesada, J. R., and Rosenblum, M.,** Partially Pure Human Immune (Gamma) Interferon: A Phase I Pharmacological Study in Cancer Patients Presented at the 3rd Annu. Int. Cong. for Interferon Res. Miami, Fla., November, 1982.

39. **Hill, N., Khan, A., Loeb, E., Pardue, A., Aleman, C., Dorn, G., and Hill, J. M.,** Clinical trials of high dose human leukocyte interferon, in *Interferon: properties and clinical uses,* 667-677, Wadley Institutes Press, Dallas, 1979.

40. **Gidlund, M., Orn, A., Wigzell, H., Senik, A., and Gresser, I.,** Enhanced NK cell activity in mice injected with interferon and interferon inducers, *Nature,* 273, 759, 1978.

41. **Herberman, R. B., Djeu, J. Y., and Kay, H. D., et al,** Natural killer cells: characteristics and regulation of activity, *Immunol. Rev.,* 44, 43, 1979.

42. **Herberman, R. B., Ortaldo, J. B., and Bonnard, G. D.,** Augmentation by interferon of human natural and antibody-dependent cell-mediated cytotoxicity, *Nature,* 227, 221, 1979.

43. **Zarling, J. M., Eskra, L., Borden, E., Horoszewiez, J. S., and Carter, W.,** Activation of human natural killer cells cytotoxic for human leukemia cells by purified interferon, *J. Immunol,* 123, 63, 1979.

44. **Koren, H., Anderson, A., Fischer, D., Copeland, C., and Jensen, P.,** Regulation of human natural killing. I. The role of monocytes, interferon, and prostaglandins, *J. Immunol.,* 127, 2007, 1981.

45. **Khan, A., Martin, E., Webb, K., Weldon, D., Hill, N., Duvall, J., and Hill, J.,** Regression of malignant melanoma in a dog by local injection of a partially purified preparation containing human alpha lymphotoxin, *Proc. Soc. Exp. Biol. Med.,* 169, 291, 1982.

46. **Khan, A., Hill, N., Ridgway, H., and Webb, K.,** Preclinical and phase I clinical trials with lymphotoxin, in *Human Lymphokines: The Biological Immune Response Modifiers,* Academic Press, 1982, 621.

47. **Welte, K., Wang, C., Martelsman, R., Venuta, S., Feldman, S., and Moore, M.,** Purification to Homogeneity of Human Interleukin 2 (IL2) and its Molecular Heterogeneity, *Proc. 13th Intl. Cancer Cong.,* 655, 1982.

48. **Borgström, S., vonEyben, F., Flodgren, P., and Sjögren, O.,** Regression of Metastatic Malignant Melanoma Induced by Combined Treatment with Alpha Interferon and Cimetidine, *Proc. 13th Intl. Cancer Cong.,* 514, 1982.

49. **Hill, N., Pardue, A., Khan, A., Aleman, C., Hilario, R., Osther, K., and Hill, J.,** IFN-alpha and Cimetidine Combinations in Malignant Melanoma, presented before the 3rd Ann. Intl. Cong. for Interferon Res., Miami, Fla., November, 1982.

50. **Burtin, C., Scheinmann, P., Salomon, J. C., and Lespinats, G.,** Cimetidine, the immune system, and cancer, *Lancet,* 1, 1900, 1981.

51. **Meretey, K., Room, G., and Maini, R. N.,** Effect of histamine on the mitogenic response of human lymphocytes and its modification by cimetidine and levamisole, *Agents and Actions,* 11, 84, 1981.

52. **Lang, I., Gergely, P., and Petranyi, Gy.,** Effect of histamine receptor blocking on human spontaneous lymphocyte-mediated cytotoxicity, *Scand. J. Immunol.,* 14, 573, 1981.

53. **Simon, M. R., Salberg, D. J., and Crane, S. A.,** *In vivo* cimetidine augmentation of phytohemagglutinin-induced human lumphocyte thymidine uptake, *Transplantation,* 31, 5, 400, 1981.

54. **Ogden, B. E. and Hill, H. R.,** Histamine regulates lymphocyte mitogenic responses through activation of specific H_1 and H_2 histamine receptors, *Immunology,* 41, 107, 1980.

55. **Gifford, R. R. M, Ferguson, R. M., and Voss, B. V.,** Cimetidine reduction of tumor formation in mice, *Lancet,* 1638, 1981.

56. **Osband, M. C., Shen, Y. J., Schlesinger, M., Brown, A., Hamilton, D., Cohen, E., Lavin, P. and McCaffrey, R.,** Successful tumour immunotherapy with cimetidine in mice, *Lancet,* 1, 636, 1981.

57. **Quesada, J. and Gutterman, J.,** Antitumor Effects of Partially Purified Human Alpha-IFN in Renal Cell Carcinoma, presented before the 3rd Ann. Intl. Cong. for Interferon Res., Miami, Fla., November, 1982.

58. **Carter, S. and Wasserman, T.,** The chemotherapy of urologic cancer, *Cancer,* 36, 729, 1975.

Chapter 5

PRECLINICAL ASSESSMENT OF BIOLOGICAL PROPERTIES OF RECOMBINANT DNA DERIVED HUMAN INTERFERONS

Nowell Stebbing and Phillip K. Weck

TABLE OF CONTENTS

I. INTRODUCTION

The discovery of pharmacologically active agents produced naturally by tissues, such as insulin, ACTH, growth hormone, and thyroxine, has generally led to their rapid clinical use. Circulating interferon titers have been found after viral infections in man and animals[1,2] indicating that they may naturally limit virus infections. However, clinical use of interferons has been slow because of difficulties in obtaining large amounts of materials and inadequate data relating to treatment regimens and target diseases. Use of cell culture assays originally indicated that interferons are soluble proteins and are mediators of antiviral effects against many viral infections in target cell cultures. Interferons have now been shown to have numerous other effects in various cellular assays, particularly relating to immune functions. Thus, the relation of the antiviral bioassay to antiviral effects in vivo has become less obvious. Various interferon preparations have already been used clinically and these studies have been extended primarily to neoplastic diseases. The initial rationale for use of interferons against neoplasms related to the supposed viral etiology for some of these diseases.[3] However, other explanations for antitumor effects are now apparent. The very varied biological properties of interferons and the recognition of a range of proteins with interferon activity necessitates reappraisal of the potential clinical uses of interferons. Elucidation of structural features of interferons now offers the possibility of relating structure to the various biological properties and consequently the possibility of designing interferon analogs with specific desired properties. These developments are dependent on experimental investigation of the role of the various properties of interferons in limiting viral and neoplastic diseases. Appropriate preclinical studies in animals are necessary to achieve better understanding of in vivo properties of interferons of primary importance for efficacy and should indicate properties warranting detailed in vitro analysis in relation to in vivo effects.

Despite their discovery over 20 years ago and their remarkable biological potency, knowledge of the primary amino acid sequence and structure of interferons is recent and progressed dramatically with the application of recombinant DNA techniques.[4] A wide range of activities has been ascribed to interferons but definitive assessment of properties has depended on the recent production of highly purified and homogeneous preparations. Recombinant DNA techniques have led not only to determination of the structure of human interferons but also production of individual molecular species of interferon free from other species and other proteins simultaneously induced in human cell cultures. In this way many of the varied biological properties of interferons, previously determined with relatively crude preparations, have been confirmed and extended using defined preparations. Comparison of the properties of different molecular species of IFN-α and molecular hybrids constructed from various species, has indicated the extent to which the biological properties of interferons may be separated. These studies have allowed preliminary investigation of structure/activity relationships. The production of hybrid interferons from related interferon genes with common restriction enzyme sites has provided a method for producing unique analogs, thus increasing the potential for structure/activity studies.

The varied biological activities of interferons and observations showing that antiviral and antitumor effects are not related to the direct actions of interferons on cells in culture has raised basic questions regarding selection of interferons for clinical development. Analysis of in vivo effects clearly requires understanding of the mechanisms involved and the contribution of those properties measured by the various in vitro bioassays. There is a need to understand the relation between antiviral and antitumor effects and the relation between in vitro and in vivo activities. Other reviews have considered current clinical studies. Biochemical and other effects of interferons, which seem of less immediate importance for development of interferons as clinical therapeutic agents have been considered elsewhere and are not covered in the present review. These reviews, rather than primary references, are cited where appropriate. Recent data, particularly those arising from use of recombinant DNA derived materials, are here reviewed with reference to mechanisms of potential significance for clinical development of these interferons.

II. STRUCTURE OF HUMAN INTERFERONS

Interferons are classified in three major groups, IFN-α, -β, and -γ, depending on biological and physicochemical features. IFN-αs are produced in leukocytes and lymphoblastoid cells by various inducers, including virus infections. IFN-β is produced in fibroblast cells by similar stimuli and, like the IFNαs, is stable at pH 2. It is now apparent that IFN-γ is produced primarily by a particular subset of T lymphocytes in response to mitogens and antigens[5] and is labile at pH 2. The major groups of interferons are antigenically distinct because antisera to each class of interferon do not neutralize interferons of the other classes. A list of recombinant DNA derived human interferons is shown in Table 1.

The amino acid sequences of 13 distinct human leukocyte interferons (IFN-αs) have been determined by molecular cloning and these consist of 165 or 166 amino acid residues.[6-9] Hybridization of DNA probes to the human genome indicates the existence of about 16 distinct genetic loci for leukocyte interferons.[8,10] Each IFN-α subtype differs from other subtypes in about 8 to 29 amino acid residues and the overall homology between all the known IFN-α sequences is 52%. A cor.sensus sequence for these IFN-α subtypes is shown in Figure 1. All but one of the IFN-αs contain 4 cysteine residues (IFN-αD contains 5) and the two disulphide bridges shown in Figure 1 are known to exist in IFN-αA, whose structure has been most extensively studied.[11,12] IFN-αA contains 165 residues (residue 44 is taken as a gap in the comparative data summarized in Figure 1). Molecular heterogeneity in natural IFN-α preparations from CML or lymphoblastoid cells has also been detected.[13-15] In addition to the distinct subtypes of IFN-α described, variant forms of several subtypes have been detected which differ in only 1 to 4 amino acid residues, and these differences arise from one or two differences in the nucleotide sequences. These variants and the known relationships between related IFN-αs cloned in different laboratories, are listed in Table 1. The variants may represent allelic forms of the same interferon subtypes and represent a degree of natural polymorphism. That these closely related subtypes are allelic forms is supported by close similarities in the signal peptide regions. For IFN-α_{4a} and -α_{4b} comparison of the chromosomal DNA segments shows identity for regions in excess of 3 kb, in both the 3' and 5' flanking regions.[10] These data strongly support the notion that these two related subtypes are allelic variants.

Only one IFN-β (IFN-β_1) has been detected by recombinant DNA methods and only one gene has been identified by hybridization to the human genome.[8] IFN-β has been cloned independently four times and the same structure identified.[16-19] The cloned fibroblast interferon contains 166 amino acid residues and shows some homology with the IFN-αs: 38/166 amino acid positions are common between IFN-β_1 and all the known IFN-αs. Evidence for additional IFN-β genes comes from production in mouse/human hybrid cells of oocyte translatable mRNA which does not hybridize with an IFN-β_1 probe and is translated into an interferon neutralized only by antibody to human IFN-β.[20] There are reports of two size classes of mRNA coding for fibroblast interferon.[21,22] The smaller species codes for IFN-β_1 and the larger species is translatable into proteins with IFN-β activity but the two IFN-β mRNA classes do not cross-hybridize.[21] However, clones of the second class of putative IFN-β species have not been sequenced at the DNA level or expressed in bacteria and shown to be distinct in amino acid sequence.

Human immune interferon (IFN-γ) consists of 146 amino acid residues with no extensive homology with IFN-αs or IFN-β and there appears to be only a single gene for IFN-γ which contains a 20 amino acid signal sequence.[23,24] Molecular heterogeneity of natural IFN-γ

Table 1
CLONED FORMS OF HUMAN INTERFERONS

Interferon class			Molecular designation	Amino acid differences between variant subtypes	Ref.
Leukocyte, IFN-α	Group I	IFN-α_1, IFN-αD	Position 114 ala, val	7, 8
		IFN-α_2, IFN-αA	Position 23 arg, lys	8, 148
		IFN-α_5, IFN-αG	Identical	8, 35
		IFN-α_6, IFN-αK	Identical	35
		IFN-α_{4a}, IFN-α_{4h}	Positions 51 and 114 ala, glu or thr, val	35
	Group II	IFN-α_7, IFN-J	Identical	35, 46
		IFN-αC		8
		IFN-αC$_1$	Differs from -αC in 10 residues	9
		IFN-αI		39
		IFN-αL		46
	Intermediate	IFN-αF		8
		IFN-αH, IFN-$\alpha_{\lambda 2h}$	Identical (H$_1$ differs in 4 residues)	8, 35
		IFN-αB, IFN-αB$_2$	Position 98—101 val, leu, cys, asp or ser, cys, val, met	8, 60
		IFN-α_8	Identical to IFN-αB$_2$	35
Fibroblast, IFN-β		IFN-β_1		16—19
Immune, IFN-γ		IFN-γ_1	Position 140 gln or arg	23

Note: Amino acid positions are numbered from the N- to the C-terminus using the scheme shown in Figure 1 for IFN-αs. For variants of the same form or subtype, the order of amino acid differences is the same as the order in which the molecular forms are shown. Grouping of the IFN-αs is according to the scheme of Weissmann[35] discussed in Section II.

preparations indicates that there may be additional forms of this interferon. In part the observed heterogeneity may be due to various degrees of glycosylation because there are two potential glycosylation sites, at amino acid residues 28 and 100. However, there is also evidence for other physicochemical differences between components of natural IFN-γ preparations.[25] For this reason it seems prudent to refer to the one cloned form of IFN-γ identified so far as IFN-γ_1 in conformity with the designation of the cloned fibroblast interferon.

It is apparent that interferons induced in continuous cell cultures by viral infections consist of a mixture of subtypes with distinct properties. Allen and Fantes[14] first raised this possibility with data indicating a family of human lymphoblastoid interferons. It was evident from previous work that lymphoblastoid interferon contained predominantly leukocyte interferons and approximately 15% fibroblast interferon.[26,27] Subsequently, lymphoblastoid interferon activity has been separated into major peaks by G-75 Sephadex® column chromatography. The majority of these proteins have molecular weights between 17,500 and 22,000.[14,28] Polyacrylamide gel analyses have shown the presence of seven subtypes in this molecular weight range and an additional band at 26,000 MW. All these fractions have antiviral activity. However, the second major peak of interferon is more active on the bovine MDBK cells than the first peak. The generation of a monoclonal antibody (NK-2) to lymphoblastoid interferon has allowed further separation of these eight subtypes by their differential ability to bind to an antibody column.[29] The molecules that do not bind to this column show significant antiviral acitivity on mouse cells and are more active on bovine cell than human cells.[30] Cloning of cDNA from induced namalva cells confirmed the presence of IFN-β, and indicated that IFN-αA may be a dominant IFN-α subtype in lymphoblastoid interferon preparations.[31]

FIGURE 1. Consensus Sequence of Human Leukocyte Interferons. The conventional single-letter code is used to indicate amino acids. Residues in squares are common to all known IFN-α subtypes. The disulphide bridges are for IFN-αA.[11] IFN-α₂ = IFN-αA has only 165 residues and position 44 is therefore omitted for this interferon in order to optimize homology. The cys at position 99 occurs at position 100 in the case of IFN-αB.

Interferons induced in CML cells are also a complex mixture of molecules and have both antiviral and antiproliferative activity in highly purified form.[13] These biological properties co-purify but the ratio of antiviral to antigrowth activity is not consistent among all species. In addition, the ratio of activity on MDBK and human cells also varies.[32] A monoclonal antibody has been found to separate natural IFN-α into two classes of IFN-α subtypes.[33] A different monoclonal antibody to natural IFN-α, used on an affinity column, has been found to bind more leukocyte interferons than lymphoblastoid interferons and this column appears to have varying affinity for different IFN-α fractions.[34] Additional information on the sequence of subtypes isolated from natural IFN-α preprations and their biological activity would add a great deal to understanding their relationship to the multiple subtypes of human alpha interferons obtained by cloning in *E. coli*.

Examination of the amino acid sequences of the cloned IFN-αs shows a high degree of positions conserved in all subtypes, as indicated in Figure 1. In some of the variable positions there are only two alternative amino acids to be found in the known IFN-α subtypes. Thus postion 16 is either met or ile. Several subtypes with the same amino acid residue at position 16 also have identical residues at other positions, principally positions 14, 71, 78, 79, 83, 154, and 160.[35] Based on these similarities the IFN-α subtypes can be classified into groups as shown in Table 1. Three IFN-α subtypes have features of both groups I and II and are listed as intermediate in Table 1. Of these IFN-αH and IFN-αF appear to be natural hybrids: IFN-αH is like group I up to residue 79 and then like group II subtypes. Overall IFN-αF is like group II subtypes with features of group I around residue 71 and at the carboxyterminus. The IFN-α₈ or IFN-αBs are sufficiently distinct that they may represent a third group of IFN-αs. Consideration of similarities at the gene sequence level indicates that the IFN-α gene family arose at least 33 million years ago with the possibility of an early divergence into two major groups.[35]

Although natural IFN-α preparations are mixtures of various subtypes it seems that different cells and induction procedures result in different ratios of the various subtypes. From the frequency of bacterial transformants obtained with cDNA from the human myeloblastoid cell line, KG-1, IFN-αA and -αD appear to be the predominant species, comprising almost 80% of the total species present.[8] From N-terminal amino acid sequence data the major species in lymphoblastoid interferon would now appear to correspond to IFN-αB, -αF or -αI.[36] The other known IFN-α subtypes are eliminated but it remains possible that the sequence determined is really the average of the several subtypes present in the lymphoblastoid interferon preparation. However, the sequence data show that lymphoblastoid interferon must contain little or none of the group I IFN-α subtypes. In contrast, positions 14 and 16 in the N-terminal sequence data for CML leukocyte interferon[37] shows that it consists predominantly of group I subtypes. The three major species of interferon from CML cells comprise approximately 50% of the total interferon species.[38] Although these species lack the 10 C-terminal residues predicted from the cloned materials, the sequences correspond to IFN-αA, except for residue 11.[38] It is not clear whether or not all the genes for the IFN-α subtypes in any one individual are expressed with different induction procedures.

Analysis of the genetic organization of IFN-α genes indicates that several of the genes are closely linked and that they lack introns.[7,9,10,39] These genes have all been assigned to human chromosome 9.[40,41] IFN-β$_1$ also lacks introns[42-44] and is also located on chromosome 9.[40] Several of the IFN-α genes are found in the same or overlapping genomic DNA clones.[9,10,45,46] Human fibroblasts with reciprocal translocations involving chromosome 9 have been used to locate the IFN-α and IFN-β$_1$ genes in a cluster on the short arm of chromosome 9.[47] At least 5 pseudogenes, which cannot be translated into functional interferon products because of the presence of deletions and stop codons have also been detected for IFN-α.[10,46] In addition, one transcribed gene, IFN-αE, isolated by reverse transcription of mRNA, also contains several stop codons.[8] Although IFN-αE may be referred to as a pseudogene, it differs from other pseudogenes, such as those of the globin family,[48] because cDNA cloning demonstrates that it is transcribed. It is possible that IFN-αL (see Table 1) is not transcribed because the gene contains a stop codon in the leader sequence. Of the IFN-αs listed in Table 1 some have only been identified at the gene level and they may not be expressed because of genetic features in flanking regions. These IFN-α subtypes are: IFN-α$_{4a}$, -α$_{4b}$, -αC$_1$, -αI, -αJ and -αL. Unlike the IFN-β and IFN-α genes, IFN-γ contains introns.[49]

The frequency at which bacteria are transformed by cDNA derived from the mRNA of induced human cell cultures might be expected to reflect the frequency of the mRNA species for different subtypes and hence the relative abundance of the various subtypes in natural preparations, assuming that translation is proportional to mRNA frequency. This may not be the case. The DNAs coding for different proteins may differ in their transformation ability in bacteria. This possibility is consistent with the observation that the IFN-αA and -αD subtypes occur most frequently on cloning cDNA from induced leukocytes[8] but a monoclonal antibody (NK-2), against a natural lymphoblastoid interferon preparation, which binds and neutralizes IFN-αA and -αD does not neutralize natural leukocyte interferon preparations.[281] The difference between frequency of DNA transformation and estimated relative abundance of interferon subtypes may reflect an unexpected effect of DNA structure on transformation frequency and this may prove important in cloning particular interferon subtypes and other human genes. In the case of cloning in mammalian cells, a difference in genotypic transformation frequency has been demonstrated for different genes in different cell lines.[50]

The human IFN-αs do not seem to be glycosylated in their natural form[14] and no known glycosylation sites occur in the human IFN-α species so far identified.[8,35,39,40] IFN-β$_1$ contains one known glycosylation site and naturally appears to contain approximately 18% carbohydrate.[51] In its natural form IFN-γ appears to be glycosylated and there are two known glycosylation sites.[23] IFN-β contains only three cysteines: there is no cysteine near the N-terminus, as occurs for all the IFN-αs. However, it is likely that IFN-β contains a disulphide bridge which is essential for antiviral activity. Changing the cysteine at position 141 to tyrosine destroys antiviral activity.[52] Whether other biological properties of IFN-β$_1$ are thus lost is unknown. IFN-γ contains two cysteines, at positions one and three and these do not seem to be involved in any intramolecular structure. Despite absence of glycosylation when expressed in bacteria, cloned human IFN-β$_1$ and IFN-γ$_1$ have biological properties comparable to those of the natural interferon preparations in all aspects studied to date.

The homology between IFN-αs at the nucleotide level is even greater than at the amino acid level and common restriction enzyme sites occur in the genes for the various IFN-α subtypes. A common *Bgl*II site occurs in the genes for IFN-αA and -αD at a position corresponding to amino acid residue 61 and a *Pvu*II site occurs in the genes for several IFN-α subtypes at a position corresponding to amino acid residue 91. These common restriction enzyme sites have allowed construction of hybrids such as IFN-αAD(Bg1): the first subtype indicated alphabetically is the source of the N-terminal fragment and the second, the source of the C-terminal fragment and the site of recombination is designated by the abbreviation for the restriction site involved. In addition to the IFN-αA and -αD hybrids a series of hybrids have been formed involving a common *Pvu*II of other IFN-α genes (see Section VI).

III. BIOLOGICAL PROPERTIES OF HUMAN INTERFERONS

A cloned human leukocyte interferon, IFN-α$_1$, derived from *E. coli*, has been shown to have a range of biological activities attributed to impure natural preparations including NK-cell stimulation, enhanced antibody-dependent cell-mediated cytotoxicity, suppression of leukocyte migration as well as antiviral and antiproliferative effects.[53] Although these studies involved an impure preparation, subsequent studies with various purified cloned interferons have confirmed that these and other effects ascribed to natural interferon preparations occur with the cloned materials.[54-57] These studies involving individual molecular species of interferons free of the other human proteins occurring as contaminants in natural preparations, clearly demonstrate that the various properties of interferons reside in the same molecular entities. Although individual molecular subtypes of IFN-α appear to have distinct properties, overall the properties of purified cloned IFN-α subtypes are comparable to those ascribed to natural preparations.

In addition to their antiviral and antitumor effects, interferons have also been found to modulate a number of cellular and humoral immune responses and at the biochemical level several effects leading to inhibition of protein synthesis have been observed. Direct antiviral and antiproliferative effects in cell cultures are associated with inhibition of protein synthesis by stimulation of a kinase known to phosphorylate a 67K protein and initiation factor eIF-2 or stimulation of 2'-5'A production leading to activation of a nuclease which degrades mRNA.[58] However, the relation, if any, between these direct biochemical mechanisms and the indirect effects of interferons remains obscure. Without doubt some of the biological properties of interferons are, in part at least, independent while others are secondary to primary effects, perhaps on protein synthesis or cell membranes. Although the independent effects of interferons may not be due entirely to distinct molecular domains it should be possible to construct derivative proteins which retain some of the biological properties and lack others. Successful screening of derivative interferons for desired properties will be facilitated by understanding which properties may be inherently linked and those that may be independent. Studies with recombinant DNA derived materials have indicated effects which are independent and some which may be dependent. Identification of cell lines lacking particular biochemical properties has allowed determination of features which are associated with or necessary for particular biological effects.

A. Antiviral and Antiproliferative Effects in Cell Cultures

Individual leukocyte interferon subtypes show distinct antiviral effects against different viruses in various mammalian cell cultures.[53,59-63] Hybrid interferons derived from these subtypes also have distinct antiviral activity and in the case of IFN-αA and -αD, mixtures have additive effects, whereas the -αAD hybrids have greater activity than the parental subtypes and the -αDA hybrids generally have much lower activity.[64] These observations have been confirmed using highly purified materials and a notable feature of IFN-αAD(Bg1) is its high activity in mouse cells[64] and against infections of mice.[65] IFN-αAD(Bg1) also shows activity against the L1210 leukemia in BDF$_1$ mice and these properties have allowed investigation of antiviral and antitumor effects in rodents.[61,66] However, even the individual IFN-α subtypes, obtained by cloning genes, have properties not apparent from the natural mixtures of IFN-α.[59,64] This indicates that the natural subtypes may be modified in some way or interactions occur between the subtypes. No such interactions modifying activity in different cell lines or against different viruses have been observed with IFN-αA and IFN-αD.[64] However, it is noteworthy that protease treatment of a natural leukocyte IFN-α preparation can alter species specificity and some active fragments are slightly smaller than the starting material.[67]

Studies of antiproliferative effects of interferons are more recent than antiviral effects and the data for cell culture studies are more difficult to relate to treatment of disease. However, many human tumor cell lines are inhibited by interferon in cell cultures as are tumor stem cells from freshly explanted tumors.[68] Various natural subtypes and hybrid human IFN-αs, produced in *E. coli*, have also shown inhibition of a range of tumor stem cell cultures.[69] In this study IFN-αA inhibited more tumor stem cell cultures than did IFN-αD. Interestingly, the tumors which were most sensitive to the cloned interferons were also those that were most sensitive to a range of cytotoxic drugs.[69] A similar correlation between sensitivity to IFN-α and chemotherapeutic agents was noted earlier for stem cell cultures from ovarian carcinomas.[70] The proliferation of normal cells is also inhibited by interferons but, overall, transformed cell lines and tumor stem cells appear to be inhibited to a greater extent than normal cells and hematopoeitic cells appear to be more sensitive to IFN-αs whereas most other cells are more sensitive to IFN-β.[68] Although there is, overall, a correlation between antiviral and anticellular effects of interferons, some of the exceptions could provide useful analytical systems for determining differences in mechanisms. Thus human IFN-αs that have potent antiviral activity against various bovine cell cultures show no antiproliferative effects against these cells.[71] Comparison of various IFN-α subtypes in cervical epithelial cells indicates that those with highest antiviral activity have lowest antiproliferative activity and vice versa.[62]

Cell culture studies have demonstrated that individual subtypes of IFN-α from *E. coli* are as effective as natural preparations in inhibiting growth of transformed cells.[54,72] It is apparent that the predictability of cell culture effects for clinical efficacy is of considerable importance. Too few studies have been carried out to really test such correlations and studies with IFN-γ are particularly limited. However, the antiproliferative activity of human IFN-γ appears to be greater than that of IFN-α or IFN-β.[73] Differentiated or functionally active cells are generally more sensitive to the antiviral effects of interferons than undifferentiated or functionally inactive cells.[74,75] The apparent correlation between antiproliferative effects of IFN-α and -β with antiproliferative effects of chemotherapeutic agents will also be of clinical importance. One indication of this observation would seem to be that tumors refractory to chemotherapeutic agents may not be better candidates for interferon therapy than tumors more susceptible to chemotherapy. However, this may not be a meaningful concern if interferons are generally utilized in regimens also involving chemotherapeutic agents. It is important to consider the efficacy of combination therapies which may be more than additive (see Section IV).

B. Cell Receptors for Interferons

Highly purified and iodinated mouse interferon has been used to determine binding to mouse cells and these studies demonstrate high affinity with a dissociation constant of the order of $10^{-11}M$.[76] Studies with human cells and natural human lymphoblastoid interferon have shown a comparable dissociation constant and use of an iodinated purified recombinant DNA derived interferon, IFN-αA, has given similar results.[77,78] Competition studies indicate that other human IFN-αs and IFN-β bind to the same extent as unlabeled IFN-α. Studies with IFN-αA labelled with ^{35}S-methionine gave similar results, indicating that iodination does not cause modification of the interferon so as to affect binding.[79] The number of binding sites per cell was estimated to be 300 for Daudi cells and about 1100 for bovine MDBK cells which are known to be sensitive to both natural and bacteria derived human IFN-αs.

Binding of interferons to cellular receptors is likely to be a critical factor determining all subsequent biological effects. However, events leading to antiviral or antiproliferative effects cannot simply be sequelae to receptor binding because different IFN-α subtypes have different activities against the same viral infection of the same cell types.[59,63,64] Presumably other molecular domains result in specific effects leading to the different properties of the different interferons and receptor binding must be distinct from activation of mechanisms leading to biological effects. Studies involving interferons from different species competing for receptors on the same cells and differential neutralization of their activity by various antibody preparations together with consideration of dose response curves gave early indications that binding and activation are distinct events.[80,81] The nature of sites on interferons involved in binding and/or activation is not clear. However, studies with molecular hybrids of human IFN-α subtypes active on mouse cells indicate that both the N- and C-halves of the interferons are required for receptor binding.[82] Various synthetic segments of IFN-α and fragments of IFN-α_1 did not show competition for binding.[78]

On human Daudi cells and bovine (MDBK) cells, various human IFN-α and -β preparations show equal competition for receptor sites.[77-79] However, on MDBK cells IFN-β competes much less effectively than various IFN-αs for the receptor identified by the labelled IFN-α (either mammalian cell lymphoblastoid IFN or IFN-αA).[78,79] Thus IFN receptors can distinguish between different interferon species. In the case of MDBK cells the lower binding affinity for IFN-β correlated with its lower antiviral activity on MDBK cells, indicating that receptor affinity may play an important role in overall activity of interferons. On human cells, IFN-γ does not show any competition for the receptors which bind IFN-αs and IFN-β.

In the case of mouse IFN-α, receptors are not found on L1210R cells, which are resistant to the effects of IFN-α and IFN-β preparations.[76] Thus in this case there is also a correlation between receptor binding and efficacy. Murine IFN-γ has antiproliferative effects on L1210R cells, indicating that IFN-γ binds to independent receptors.[75,83] However, labelled mouse IFN-α binds with comparable affinity to murine embryonal carcinoma cells which are resistant to the interferon indicating that binding alone is not sufficient for biological effects.[75] The murine embryonal carcinoma cells can be induced to differentiate and they are then susceptible to interferon.[75] However, the number and affinity of the interferon receptors are unaltered on differentiation.[75] These studies indicated that resistance of the undifferentiated carcinoma cells is not due to complete dissociation of binding and activation of interferon mediated effects because 2'-5'A synthetase activity is increased on treatment of the resistant cells with interferon.[75] Thus, the correlation between interferon binding and biochemical changes in the cell membrane and cytoplasm remains unclear. Studies using the L1210R cells demonstrated that these cells lack fatty acid cyclooxygenase activity.[84] In addition, inhibitors of this enzyme activity prevented the induction of the antiviral state by interferon.[85] Clearly, further studies on the number and properties of interferon receptors on normal and neoplastic human cells could be useful for assessing the clinical potential of various interferon preparations in particular diseases.

C. Cell Membrane Effects of Interferons

Evidence for marked changes in cell surface properties after treatment with interferons has arisen from numerous studies and these effects include changes in membrane composition and functions such as ion transport, physiological activity of specialized cells and cell mobility.[68,74,86] The relationship of these cell membrane effects to the antiviral and antiproliferative effects of interferons remains unclear.

Because the degree of saturation of fatty acids in phospholipid molecules and cholesterol content regulate membrane fluidity, interferon-induced changes in fatty acid chains may explain effects on membrane fluidity.[86] Such changes in the lipid bilayer of the cell membrane may be one of the primary reasons for the reported multiple effects of interferons on membrane structure and function. Certainly major changes in the cytoskeletal structure of cells occur following IFN treatment[87] and the rigidity of the plasma membrane is markedly increased. Human IFN-β causes an increase in cell volume and surface area and this is associated with increased numbers of concanavalin A binding sites.[88] These changes could explain the effects of IFN on the redistribution and surface density of cell surface components.[89]

Presumably the effects of interferons on structural integrity of the plasma membrane are related to the alterations in functional capacities. For example, natural IFN preparations inhibit transport of ^{3}H-thymidine[90,91] but enhance the expression of cell surface markers such as β_2 microglobulin and HLA antigens.[92,93] A recent study has shown that purified natural and recombinant DNA derived IFN-γ have greater effects on induction of HLA synthesis than IFN-α or IFN-β and that this is associated with increased synthesis of the respective mRNAs.[94] Other manifestations of IFN effects on cell surfaces include changes in the net surface charge and membrane conductivity.[95,96] If these physiological effects occur in vivo the clinical consequences could be considerable. Whether these membrane-associated alterations are related to the multiple side effects observed in Phase I clinical trials is unknown. However, it is tempting to speculate that some of the neurotoxicity observed could be related to alterations in membrane fluidity and function.

Recent evidence indicates that binding of interferon to cell receptors is species-specific and that receptors can be detected on erythrocytes and fibroblasts.[97] It is also apparent that IFN receptors are between microvilli of the cells whereas the Diphtheria toxin receptors are on the surface of the microvilli. The binding of IFN to its receptors alters the binding of this toxin, thus reducing its biological activity.[97] It is possible that other apparent interactions between interferons and other drugs, considered below, arise from redistribution or interactions between the relevant receptors induced by cell surface changes.

Blalock and co-workers have reported that interferons can have activity corresponding to specific hormones on the target cells for the hormone under study. Mouse interferon stimulates beat frequency of mouse myocardial cells[98] and causes a steroidogenic response in mouse adrenal tumor cells.[99] The antiviral activity of human IFN-α preparations has been shown to be neutralized by antiserum to ACTH.[100] This effect did not occur with human IFN-β and increasing concentrations of ACTH blocked neutralization of IFN-α by the antiserum. Similar results were obtained for natural human IFN-α using antiserum to γ-endorphin.[101] The natural human IFN-α, but not IFN-β or IFN-γ, was shown to bind to opiate receptors and cause potent endorphin-like opioid effects when injected intracerebrally to mice.[101] These effects have been interpreted as indicating structural similarity or association of the relevant hormones with IFN-α preparations.[101] Such studies must be interpreted with some caution because ACTH activity was demonstrable only after pepsin or acid treatment of the interferon preparation.[12] Subsequently, it has been shown that there are no sequence homologies between bacteria-derived IFN-αA, ACTH and γ-endorphin.[103] Biological studies have demonstrated that these hormones do not induce an antiviral state in cells and that antisera to ACTH or endorphins do not neutralize the interferons' activity.

Additional studies by Wetzel et al.[104] have shown that neither IFN-αA nor -αD activity is affected by large molar excesses of ACTH. In these experiments, antisera raised to ACTH had no effect on these two interferons and did not cross-react with the proteolytic digests of the two interferon molecules. IFN-αA, α-D and pepsin digests of these interferons failed to induce steroidogenesis in cell cultures. Although interferons apparently do not induce such effects directly, it is possible that they arise from interactions between the receptors involved after binding of the hormones or interferons and the cell surface effects of interferon may augment such interactions. Clearly the possibility of such effects in clinical situations requires careful consideration, particularly in pathological states involving hormones and in patients receiving hormone treatments.

D. Effects on Functions of the Immune System

Interferons have been shown to have profound effects on a number of different immune responses as measured both in vitro and in vivo.[105-107] Earlier studies were clouded by the fact that impure preparations of interferons containing other lymphokines were employed. However, with the advent of new purification methods as well as recombinant DNA derived materials, highly purified materials have been used to investigate immunomodulatory effects of interferons. Studies designed to compare homogeneous preparations of IFN-α, -β and -γ will hopefully lead to an indication of which interferons are more potent in altering specific parameters of immune responses.

1. Natural Killer (NK) Cell Activity

Independent studies by various laboratories have demonstrated that all three major groups of natural interferons can enhance in vitro NK cell activity. Gamma interferon from induced human peripheral blood mononuclear cells can augment NK cell activity against K562 human leukemia cells.[108] Similar results have been obtained using purified natural IFN-β[109] and purified natural IFN-α in cell cultures.[110] Whether these increased cellular functions require stimulation of the synthesis of new polypeptides is unclear. However, the treatment of human lymphocytes with either highly purified lymphoblastoid or fibroblast interferons results in the synthesis of specific peptides in the T-cell population.[111] This alteration in protein synthesis can also be achieved by the addition of bacteria-derived IFN-α_2.[112]

There are now several reports concerning the effects of recombinant DNA derived human IFN-αs on in vitro NK cell activity.[53-56,113] Herberman and his colleagues[55] have demonstrated that IFN-αA is species-specific, having no activity on mouse NK cells. As detected in antiviral assays, the hybrid IFN-αAD(Bgl) is distinct from the parental subtypes IFN-αA and -αD in NK cell stimulation and is a more potent stimulator.[54] Interestingly, the simultaneous addition of the two IFN-α subtypes does not result in additive effects on NK cell activity, whereas sequential addition does.[56] In these studies cloned IFN-β_1 was also active and the sequential addition of IFN-β_1 followed by IFN-αD resulted in optimal stimulation of NK cell activity. Whether correlations can be made between clinical investigations and such cell culture studies appears questionable. Recent evidence indicates that administration of IFN-α to patients initially causes enhancement of NK cell activity followed by a decreased activity for the duration of interferon therapy[114] and IFN-αA causes decreased NK cell activity in about $^1/_3$ of patients.[115] Stimulation of NK cell activity in vitro and inhibition in vivo has also been observed with IFN-β.[116]

There are several problems in interpretation of effects of interferons on NK cell activity in relation to treatment of disease that arise from the method of assay. Effector cells are generally treated with interferon and then added to target cells, often the erythroid leukemia cell line, K562. However, treatment of target cells with interferon decreases their susceptibility to activated NK cells[117,118] and in some diseases unfractionated lymphocyte preparations are less responsive than purified cell preparations or unfractionated lymphocytes from normal individuals.[119] Human IFN-γ appears to be significantly more effective than IFN-α or IFN-β in blocking NK cell mediated lysis of cells when the interferons are added to the target cells.[120] In this study the effector cells were treated with natural IFN-β. It remains unclear whether treatment of effector cells with other interferons causes comparable effects when the target cells are treated with IFN-α, IFN-β or IFN-γ. There is considerable variation in susceptibility of different target cells to NK cell mediated lysis and the circumstances of interferon concentrations at effector and target cells in vivo and the nature of the tumor cells are seen to be factors that could greatly affect NK cell mediated tumor cell lysis. Nevertheless there is considerable circumstantial evidence that NK cell activity is associated with beneficial effects of interferon in virus infections and tumors.[121] Furthermore, mouse NK cells cloned in vitro and transferred to NK-deficient mice have been shown to have a role in rejection of allogeneic bone marrow grafts and resistance to challenge with tumor cells.[122]

2. Other Cellular and Humoral Responses

Interferons can affect many aspects of the immune response and have been shown to alter both B cell[123] and T cell[124] functions as well as enhance the cytocidal activity of macrophages.[125] Addition of lymphoblastoid or natural IFN-α to human cell cultures suppresses non-specific immunoglobulin synthesis due to pokeweed mitogen, hydrocortisone or EB virus stimulation.[126,127] Leukocyte interferon also suppresses Con-A activated human T cells.[128] In contrast, the presence of natural IFN-α in cell cultures enhances antibody dependent cell-mediated cytotoxicity[110,129] and natural IFN-β enhances cytotoxic T cell responses.[109]

There are many variables which can determine the outcome of the effects of interferon on in vitro immune responses such as time of addition of IFN to cultures and IFN concentration. Addition of interferon 24 hr prior to stimulation of human cells with horse red blood cells caused significant enhancement in antibody production. However, simultaneous addition resulted in a marked suppression in the number of plaque forming cells.[130] The effect of natural human leukocyte interferon on B cell differentiation varied with interferon concentration.[123] Thus 100 U/mℓ caused an enhancement whereas 10,000 U/mℓ suppressed the response to pokeweed mitogen. Experiments performed with highly purified preparations of individual species of IFN-α from *E. coli* have demonstrated that IFN-αA, -αD and -αAD(Bgl) can significantly alter in vitro immune responses.[57] The addition of these IFN-α subtypes to human peripheral blood mononuclear cells resulted in significant suppression of the mitogenic response to PHA or PWM. This suppression required the continuous presence of interferon in the cultures and could be achieved with as little as 10 U/mℓ of IFN-αAD(Bgl) but required 300 or 800 U/mℓ of IFN-αA or -αD to achieve the same level of inhibition. These interferons also caused a substantial reduction in mixed lymphocyte reactions and, interestingly, pretreatment of stimulator cells alone was sufficient to achieve this decreased response. In contrast to results obtained with crude natural IFN-α,[130] all three cloned IFN-αs enhanced the human in vitro primary immune response to red blood cell antigen.[57] Clearly, human interferons can have multiple and diverse effects on human lymphocyte functions.

Interferons have been shown to activate normal monocytes and macrophages[131] as well as cells isolated from human tumors.[125] In the murine system optimal activation seems to involve priming by other lymphokines prior to triggering high cytotoxic activity.[132] Activation of macrophages has been shown to be associated with production of various enzymes including proteases.[133] Production of protease activity, measured as plasminogen activator, is stimulated in murine and human macrophages by treatment with homologous interferons.[134] The human interferon used in these studies was the recombinant DNA derived subtype, IFN-αA. Higher interferon doses (1000 U/mℓ) are required for activation of macrophages than are required for stimulation of NK cells and continuous exposure for over 12 hr is necessary.[134]

Studies in murine models have demonstrated that exogenous interferon causes enhancement in expression of Fc gamma receptors[135] and inhibition of the T suppressor cell population.[136]

In these animal studies, the time of administration of interferon critically affects the response observed. Although such responses have not been monitored closely in human clinical studies, in vitro studies indicate that such modifications do occur. Incubation of lymphocytes with human IFN-α inhibits the generation of allospecific suppressor T lymphocytes that normally develop during in vitro priming against allogeneic cells.[137] Fc receptor mediated activities, such as antibody-dependent cellular cytotoxicity (ADCC) are also influenced by the presence of interferon in cell cultures and preparations of human IFN-γ are potent stimulators of ADCC activity.[108]

Quantitation of Fc receptors has shown that supernatants from mixed lymphocyte reactions augment the expression of these receptors.[138] More recently, purified preparations of bacteria-derived human IFN-α, -β and -γ have been compared for their ability to induce the expression of Fc gamma receptors. Studies by Guyre and co-workers[139] have shown that neither IFN-α_2 nor IFN-β_1 derived from E. coli caused a dramatic increase in the number of IgG molecules bound to human monocytes. In contrast, the cloned human IFN-γ caused a five- to sevenfold increase in the number of receptors on HL-60 cells, U-937 cells and cultured human peripheral monocytes.[139] Such results suggest that IFN-γ may be a more potent modulator of these immune response than either IFN-α or IFN-β. The relevance of these effects to clinical responses remains to be determined.

E. Effects on Hepatic Cytochrome P-450 Metabolism

Many toxic agents and drugs are rapidly metabolized and cleared from the body through oxidation in the liver by pathways with broad substrate specificity. The principal mechanism depends on the cytochrome P-450 mono-oxygenase system and administration of interferon-inducing agents to man and experimental animals causes depression of this system.[140-143] Direct demonstration that this effect is due to interferon came with studies involving highly purified IFN-αAD(Bgl) in mice[144] and in this case an interferon preparation inactive as an antiviral agent in mice (IFN-αA) had no effect on P-450 metabolism. The studies in mice demonstrating suppression of P-450 with the heterologous human hybrid interferon, IFN-αAD(Bgl) have been confirmed using partially purified preparations of mouse IFN-α and IFN-β.[144] These data indicate that suppression of P-450 metabolism is correlated with antiviral activity at least against the lytic EMC virus. This may occur simply because of inhibition of protein synthesis by antiviral interferons and inhibition of protein synthesis will readily affect a major, rapidly turning over set of proteins such as those of the P-450 system in hepatocytes.

Suppression of P-450 metabolism in mice by IFN-αAD(Bgl) has been shown to have predictable effects on the metabolism of drugs known to undergo metabolism via the P-450 system. Thus the sleep-time of mice treated with hexabarbital is increased with prior treatments with IFN-αAD(Bgl) as expected for a delay in conversion to inactive metabolites.[145] The toxicity of acetaminophen, which is due to toxic metabolites arising from P-450 metabolism, is decreased after prior treatments with IFN-αAD(Bgl).[145] Induction and passive transfer of murine IFN-γ to mice has also been shown to suppress hepatic P-450 activity[146] and also to depress the metabolism of drugs.[147] Clearly, effects of interferons on P-450 metabolism should be considered whenever combinations of interferon and chemotherapeutic agents are being considered. Possible effects on chemotherapeutic agents are considered below (Section IV).

F. Efficacy Studies in Animals

The species specificity, particularly of IFN-β and IFN-γ, poses severe limitations on meaningful preclinical studies with these materials. Sub-human primates have been used to a limited extent, and IFN-α preparations have been most widely used in these studies. The activity of human IFN-α preparations in rabbit cells has also led to assessment of the efficacy of human IFN-α preparations in infections of rabbits, particularly herpes induced keratitis.

IFN-αA has been shown to reduce viremia and mortality in squirrel monkeys infected with EMC virus.[148] The corresponding subtype, IFN-α$_2$, from Weissmann's group, has also been shown to cause reduction in vaccinia virus lesions in the skin of rhesus monkeys.[149] Comparison of highly purified preparations of IFN-αA and -αD against EMC virus infections of squirrel monkeys at 10^6 U/kg/treatment showed comparable antiviral effects with substantial reduction or elimination of viremia and extension of these studies to lower virus doses showed that IFN-αD was significantly more effective than IFN-αA.[150] These studies also showed that the hybrid interferon, IFN-αAD(Bgl) was as effective as IFN-αD and that if treatments are delayed until 24 hr after infection of squirrel monkeys no significant antiviral effects are observed.[150]

EMC virus shows a different course of disease in the Bolivian and Guyanan strains of squirrel monkey. In both strains equal viremia results from equal infective doses of the virus. However, only 100 p.f.u. EMC virus is necessary to cause death of the Bolivian strain whereas even 100,000 p.f.u. causes less than 100% death of the Guyanan strain.[150] The Bolivian strain shows pronounced symptoms prior to death but at 100 p.f.u. EMC virus infection of the Guyanan strain is asymptomatic. Thus the comparable antiviral activity observed against EMC virus of these two strains of squirrel monkeys demonstrates that the effects of the interferons are primarily on virus replication and not on the sequalae to infection that lead to death.[150]

In the model involving EMC virus infections of squirrel monkeys both natural IFN-β and *E. coli* derived IFN-β$_1$ have been found to cause comparable reduction in viremia in the Guyanan and Bolivian strains.[51,151] These comparisons involved intravenously administered materials. Intramuscular administration of IFN-β$_1$ was found to cause antiviral effects at least as great as intravenously administered material.[151] Blood concentrations of the intramuscularly administered IFN-β$_1$ did not exceed 100 U/mℓ, as found also for natural IFN-β in *Cercopithecus* monkeys.[152] Intramuscular treatment of rhesus monkeys with 0.5×10^6 U/kg of IFN-αA or *E. coli* derived IFN-β$_1$ showed that both materials were effective although IFN-β$_1$ showed less pronounced effects.[153] In this study no circulating concentrations of IFN-β$_1$ were observed. The lower efficacy of IFN-β$_1$ compared with IFN-αA may not be due to absence of detectable circulating IFN-β$_1$. A comparison of IFN-β$_1$ with natural IFN-β has shown that comparable effects against vaccinia virus lesions in rhesus monkeys are observed.[154] These results, together with comparison of antiviral effects in cell cultures[51] indicate that the carbohydrate moiety present on natural IFN-β does not affect any known antiviral effects of this interferon. Absence of any differences in antiviral activity in cell cultures of IFN-β has been recently confirmed after deglycosylation of natural human IFN-β.[155]

Antiviral effects observed in fibroblast cultures may be largely unrelated to antiviral effects in vivo although any relations may be dependent on the nature of the virus or tumor being studied. Destruction of macrophages in mice destroys the efficacy of interferons against EMC virus infection, implying the involvement of mechanisms not operative in cell cultures.[156] This effect is less pronounced for mouse fibroblast interferon in which case the mechanisms involved in cell culture assays may play a role in vivo.[156] Clear evidence for the importance of indirect effects in vivo comes from studies with a strain of vaccinia virus insensitive to interferons in monkey cell cultures but susceptible when inoculated to form skin lesions.[157] Moreover, the cloned human interferon subtype, IFN-αA, is as effective against HSV-1 infection of the rabbit eye as highly potent buffy coat interferon preparations[158] although IFN-αA has poor activity in rabbit cell cultures.[59] Studies with highly purified materials have shown that IFN-αD is more effective than IFN-αA against HSV-1 infections of the rabbit corneum and *E. coli* derived IFN-β$_1$ is without any significant effects.[159] The greater activity of IFN-αD in this system may be related to its greater antiviral activity in rabbit cell cultures. However, IFN-β$_1$ does show antiviral activity in primary rabbit cell cultures.[51] Thus, in the case of HSV-1 infections indirect mechanisms appear to be important for in vivo efficacy in the rabbit system.[158] Curiously, however, efficacy in the rabbit eye model occurs with topical treatments but no intramuscular treatments.[158] Human cornea cells have been shown to produce predominantly IFN-β after viral infections but in these cells IFN-β and IFN-γ have little or no activity against HSV-1 although marked resistance is conferred against VSV.[160] Although studies in cornea cells may indicate the relative importance of direct effects of various interferons against HSV-1 infections, the importance of indirect effects mediated by different interferons in vivo remains unclear.

In monkeys and mice there appears to be an over all correlation between cell culture and in vivo efficacy for various interferons.[64,65,150] Nevertheless, it should be noted that in squirrel monkeys dose response studies indicate that IFN-αD is more effective against EMC virus infections than IFN-αA[150] although IFN-αA has a considerably higher specific activity against this virus in cell cultures.[59,64]

Antitumor effects in vivo occur against tumors selected for resistance in cell cultures to antiproliferative effects of interferons. This has been shown using mouse IFN-α and the hybrid interferon, IFN-αAD(Bgl) with L1210 cells.[161,162] It might be supposed that this could arise because NK-cell mediated tumor cell killing is the primary mechanism involved in tumor regression in vivo. However, IFN-αAD(Bgl) does not stimulate mouse NK cells against L1210 as target cells.[54] Studies demonstrating efficacy of IFN-αAD(Bgl) post-tumor inoculation indicate an immune mechanism and antibody dependent cell mediated cytotoxicity may be important for antitumor effects.[28]

Natural human IFN-β, IFN-α, and lymphoblastoid interferons have been shown to have antiviral effects against the oncogenic herpes viruses, saimiri, and ateles with much lower antiviral activity against neurotropic herpes viruses.[163] Antiviral effects in cell cultures with the oncogenic viruses does not require treatment of cells before infection and the continued presence of interferon is required in order to limit latent infection and transformation.[163,164] Cocultivation of infected cells with normal permissive cells results in infection of the permissive cells but this is prevented in the presence of natural human IFN-α.[164] It is possible that the mode of action of interferon in this case related to limitation of transfer of the herpes genome to non-infected cells. In owl monkeys infected with herpes saimiri human interferon (unspecified) has been shown to limit the virus-induced leukemia with treatments commencing after onset of overt disease.[165] Treatment of marmosets with human leukocyte interferons appears to have no significant effect on herpes saimiri-induced leukemia with treatments commencing before inoculation of the virus.[164] These results are not discouraging when considered in conjunction with the evidence reviewed above that interferon treatment after development of tumors may not be beneficial. Primate models for assessing antitumor effects of interferons are likely to prove of importance for assessing the antitumor properties of a range of human interferons and the oncogenic herpes viruses of primates could be useful.

Human tumors have been established in nude mice and this system might be considered to provide a suitable means for assessing in vivo antitumor effects of human interferons. However, the apparent importance of indirect mechanisms for in vivo efficacy renders questionable the usefulness of the nude mouse system. Curiously, human tumors have been found to grow more rapidly in mice treated with anti-serum to mouse interferons and tumors normally incapable of growing in nude mice have been established after such antiserum treatments.[166] These observations indicate that mouse interferon affects growth of human tumors in nude mice. Other studies have indicated that growth of human tumors in nude mice is inhibited by human IFN-α and -β preparations but not by treatments with exogenous mouse interferon.[167-169] Thus human interferons do appear to have direct antitumor effects and the antitumor effect of endogenous mouse interferon, if a real effect, is likely to be by indirect mechanisms. Kaplan and Slimmer[170] have shown that cloned human IFN-αs inhibit the growth of human astrocytomas in the nude mouse. An IFN-α subtype, IFN-α_2, appears to be comparable in potency to lymphoblastoid interferon against human breast cancers in nude mice.[71] Further investigation of the relative sensitivity of human tumors to mouse and human interferons, in cell cultures and in nude mice, should help to elucidate the importance of direct and indirect effects and differences in species specificity of the underlying mechanisms.

G. Pharmacokinetic Studies

Intravenous administration of IFN-α preparations is associated with a two-phase plasma decay curve.[171-173] In experimental animals and humans natural and bacteria derived IFN-αs have a primary phase $t_{1/2}$ of about 20 min. and a terminal phase $t_{1/2}$ of several hours.[115,174-177] Similar results have been obtained with human IFN-β given intravenously.[152,178-180] Intramuscular administration of IFN-β gives almost undetectable levels of interferon activity in the circulatory system.[172,178,180] It has been suggested that the low levels following intramuscular injection are due to local breakdown of IFN-β.[172] Alternatively, it is possible that muscle tissues have high numbers of receptor sites for interferons, especially IFN-β. Neither IFN-α nor IFN-β show a propensity for binding to particular tissues or pooling in particular organs.[181] The differences in pharmacokinetics between IFN-αs and IFN-β have raised questions about the role of carbohydrate moieties in determining clearance rates for interferons. Studies in which carbohydrate was thought to be removed enzymatically from human interferon demonstrated that the clearance rate in rabbits is unaffected.[182] However, these observations are now readily explained by evidence that human IFN-αs lack carbohydrate moieties.[8,14] Later studies with lymphoblastoid interferon showed that asialo-IFN was cleared more rapidly than untreated interferon but that if 85% or more of the total carbohydrate was removed, clearance rates were similar.[183] Interpretation of these data not easy in view of the observation that only the minor IFN-β component of lymphoblastoid interferon is glycosylated.

The kinetics of IFN-α clearance have been substantiated by experiments utilizing *E. coli* derived IFN-α, -β and -γ preparations in squirrel monkeys.[282] Leukocyte interferon from buffy coat cells is cleared from the circulation of squirrel monkeys at essentially the same rate as IFN-αA, -αB, -αC, -αD, or -αF following intravenous administration. As reported for natural IFN-α, intramuscular injection of IFN-αA gives detectable serum levels, whereas intramuscular injection of IFN-β_1 results in little or no biologically active interferon in the serum. Interestingly, initial studies with bacteria derived IFN-γ indicate that the pharmacokinetics of this interferon is similar to that of IFN-β_1.[283] It is possible that the comparable efficacy of IFN-β after intravenous or intramuscular treatments (see Section III.F) could arise coincidentally from very different pharmacokinetics. In Cynomolgus monkeys crude human IFN-β has been shown to produce circulating concentrations of 80 U/mℓ during infusion over 3 hr of 2×10^5 U/kg.[180] A bolus injection of the same dose of IFN-β caused a fourfold higher dose which rapidly decreased. Intrathecal administration of 2×10^5 U/kg of IFN-β resulted in CSF levels of 1000 to 2000 U/mℓ with no detectable concentrations in the blood.[180] Clearly the route and method of administration of IFN-β and other interferons could have considerable influence on the efficacy in different diseases.

In humans, relatively high and prolonged serum titers of lymphoblastoid interferon can be achieved by continuous intravenous infusion over a period of 6 hr.[184] Significant levels, of 50 to 100 U/mℓ of serum, could be detected as long as 24 hr after the infusion period. Doses of interferon as great as 50×10^6 U/m^2/day could be tolerated when an infusion system was employed and could be repeated daily for up to 6 days with no more than the standard side effects.

IFN-αs appear to be metabolized in the kidneys because no detectable interferon activity occurs in the urine and nephrectomization delays plasma clearance.[173] The relation of kidney metabolism to the pharmacokinetic properties of different interferon preparations is not yet clear. The mechanism of catabolism of interferons in the kidney is also unclear but there is the possibility of high concentrations of bound interferon in kidney tissues. This possibility is significant in relation to nephrotoxicity observed with interferons (see Section IV) and also treatment of kidney tumors.

Recently the rat has been examined as a model for pharmacokinetic studies of human interferons because differences in blood concentrations of two subtypes, IFN-αA and -αD after intramuscular or intravenous administration were found to be essentially the same as in squirrel monkeys.[185] This system has been used to examine the effect of nephrectomization. In rats there is no significant difference in the clearance of IFN-αA, -αD, or the two hybrids, IFN-αAD(Bgl) and IFN-αAD(Pvu) after intravenous administration and the terminal $t_{1/2}$ values are all essentially the same.[185] As in rabbits treated with a natural human IFN-α preparation, nephrectomization of rats has a profound effect on clearance of the human IFN-α subtypes and hybrids. In all cases clearance is greatly decreased compared with sham operated control rats and there is a particularly pronounced effect on the initial clearance phase. In contrast, the clearance of IFN-β_1 is essentially unaffected by nephrectomization and there is no difference in clearance pattern between the cloned and a natural IFN-β preparation.[185] The apparent volume of distribution for the IFN-βs is closer to that of the IFN-αs in nephrectomized rats that in sham operated controls.[185]

Consideration of the pharmacokinetics of exogenously administered interferons could be usefully compared to the localization and tissue distribution of interferons induced by viral infections. In cases where detailed investigations have been made, remarkably different tissue distributions have been observed even with the same virus when infection is initiated by different routes and there is no clear relation between susceptibility of a virus in cell cultures to interferon and in vivo efficacy.[186] However, time of appearance of interferon in target organs is related to course of disease, and exogenous interferon is generally effective when the concentration achieved in the target organ is higher or occurs earlier than during infection, particularly in the case of central nervous system infections.[186] The concentration of interferons in cerebrospinal fluid after parenteral administration is generally only a few percent of that in the circulation and this has been confirmed by studies in primates.[187] The low partitioning of exogenously administered interferon between the brain and blood and the relation to circulating concentrations is not without promise for treatment of brain tumors or infections. Significant brain concentrations of IFN-β are unlikely to be achieved with intramuscular treatments but slow intravenous administration could be useful. In view of the involvement of the kidney in elimination of IFN-αs factors affecting kidney metabolism could be used to maintain prolonged circulating concentrations and hence increased and prolonged concentrations in cerebrospinal fluid.

As we have seen (Section III.F), antiviral activity is not correlated with sustained or even transient circulating concentrations of interferon because in monkeys intramuscular treatments with IFN-β are at least as effective as intravenous treatments. These observations, in combination with the existence of high affinity receptors for interferons, indicate that pharmacokinetic studies are likely to be of significance with regard to implications for interferon metabolism rather than absolute blood concentrations. Preclinical studies in primates relating pharmacokinetic parameters and metabolism of interferons to efficacy are very limited but could be of considerable value in developing interferons, particularly new analogs, for specific clinical indications. The involvement of the kidney in clearance of IFN-αs is of obvious clinical significance and the consequences of sustained blood concentrations of IFN-αs in cases of abnormal kidney function needs to be considered. That minor structural differences between interferons can have pronounced effects on pharmacokinetic properties has been established using hybrid interferons and these results are considered in Section VI.

H. In Vivo Induction and Modulating Factors

Natural induction of IFN-α and -β by virus infections is associated with a transient rise in interferon titers in various tissues. However, the hyporesponsiveness observed with synthetic interferon inducers such as poly(I). Poly(C) also seems to occur with viral infections. Various factors limiting interferon induction have been observed in virus infected and tumor bearing mice.[188,189] These factors may or may not affect the activity of exogenously administered interferons. However, factors have been identified in human serum that inactivate interferons.[190] Conditions affecting the production or activation of such factors in various pathological states are likely to be of importance for clinical use of interferons.

An inhibitor of interferon action has been found in mouse lymphokine preparations induced by means that also induce interferon.[191] Exogenously administered interferons, derived from *E. coli* will lack such factors and the activity of the cloned materials could thus be significantly different from that of human cell derived interferon preparations and interferons induced in vivo. An additional factor that may affect the efficacy of exogenously admininstered interferons relates to other natural mechanisms involved in defense against disease. Thus neutralizing antibodies and induced interferons both play a role in limiting and determining the course of disease in experimental influenza virus infections of mice.[192] Interferons directly affect antibody synthesis (Section III.D.2) and T cells are involved in both interferon and antibody production. Thus there are many possible interactions between elements of the immune system that could modulate efficacy of exogenously administered interferons. Thyroid cells appear to be particularly sensitive to interferons and interferons markedly stimulate iodide uptake.[74] Stimulation of iodide uptake appears to be unrelated to antiviral effects because it is not blocked by actinomycin D, indicating that transcriptional and translational events are not necessary.[74] Thus interferons may have direct autoregulatory or stimulatory effects on thyroid activity which affects overall immune responses.

Whereas production of IFN-α and -β in vivo is associated with viral infections, IFN-γ has not been observed naturally in vivo. IFN-γ has been routinely induced in lymphocyte cultures by mitogens or antigens causing cell division but the nature of natural inducers is uncertain. The thymic peptide, thymosin α₁, has been shown to both augment induction of IFN-γ with mitogens and alone to induce IFN-γ in the absence of mitogenesis.[193] Thus cell division per se does not seem to be essential for induction of IFN-γ. These effects occur in peripheral blood lymphocyte cultures at thymosin α₁ concentrations in the ng/mℓ range which are known to be achieved in human peripheral blood.[194] The crude thymic extract, TP-1, alone does not induce IFN-γ but it does augment IFN-γ induction by mitogens.[195] TP-1 probably contains thymosin α₁ but other peptides, such as thymosin α₇, which have immunosuppressive effects, may counteract the effects of any thymosin α₁ present. These results indicated the possibility that synergistic interactions between thymic peptides and natural mitogens such as interleukin-2 (IL-2)[196] could regulate IFN-γ production in vivo.

Induction of IFN-γ in mixed leukocyte cultures involves several cell types but the requirement for other cells for IFN-γ production by suppressor T cells can be replaced by IL-2.[197] Furthermore, the neuroendocrine hormones, arginine vasopressin, and oxytocin can replace the requirement for IL-2 in the production of IFN-γ.[198] Mitogens that cause induction of IFN-γ also induce many other lymphokines and the thymic peptide, thymosin α₁ also causes induction of several lymphokines.[199] Thus the production and activity of IFN-γ may be profoundly affected by other lymphokines and neuroendocrine hormones.

Murine and human IL-2, free of detectable interferon activity, have been shown to augment IFN-α mediated NK cell activation such that the maximal NK cell activity induced by IFN-α alone is exceeded.[200,201] Moreover, in these studies, the IL-2 preparations alone were capable of stimulating NK cell activity. Human leukocytic pyrogen or lymphocyte activating factor alone is unable to stimulate NK cell activity but in the presence of either IL-2 or IFN-α there is a marked augmentation of human NK cell activity.[201] Our knowledge of the interactive effects between lymphokines is still rudimentary but the possibility of cascade effects is already apparent[201] and their effects on the activity of interferons could be considerable.

Various genetic factors are also known to modulate the efficacy of interferons and some of these are disease specific. These data, primarily from murine systems, are considered in Section VII.D.

IV. INTERACTIONS BETWEEN INTERFERONS AND OTHER DRUGS

Relatively few studies have been devoted to investigation of combination treatments using interferons and other agents. In cell cultures no interaction was observed between methotrexate and IFN-α treatment of osteosarcoma or lymphoblastoid cells.[203] Several agents, including vinblastine, were found to prevent the antiviral and antiproliferative effects of mouse interferons in mouse cell cultures.[204] However, cytosine arabinoside and IFN-α showed additive effects on inhibition of human marrow cells except when interferon preceded cytosine arabinoside.[205] Chemotherapeutic agents or radiation therapy have been observed to be beneficial in combination with mouse interferons in mouse tumor models.[206-208]

The interaction of interferon with chemotherapeutic agents is of considerable importance in the clinic. Many patients with neoplastic disease are at risk from a number of serious viral infections. Herpes viruses, vaccinia, and cytomegalovirus infections all pose serious problems to either immunosuppressed patients or individuals on standard chemotherapy regimens. Because treatment with exogenous interferon may be very beneficial in such cases, it is important to know whether antineoplastic agents or other antiviral agents affect the antiviral activity of interferon. A positive interaction has been observed for combinations of human IFN-α and Acyclovir® against human herpes virus.[209] In vitro studies have shown that therapeutic concentrations of vincristine and hydrocortisone adversely affect the antiviral activity of either natural IFN-α or IFN-β.[210] Adriamycin and 6-mercaptopurine also reduced the interferon's activity but only when cells were pretreated with therapeutic doses of these compounds. As suggested by Cesario and colleagues,[210] such studies raise the possibility that virus infections in patient's receiving chemotherapy may be much worse as a result of reduced activity of the interferon. If interferons are to be used clinically in combination with other agents, it may be best to give the interferon at a time when the effects of other agents are minimal. Cell culture studies show that inhibitors of cyclooxygenase activity decrease efficacy of interferon.[84,85] However, anti-inflammatory drugs lacking cyclooxygenase inhibitory activity did not suppress IFN antiviral effects and in rhesus monkeys comparatively high doses of acetosalicylic acid have no deleterious effect on the activity of various cloned and natural human interferons against vaccinia virus infection.[153]

Abrogation of the efficacy of cyclophosphamide therapy occurs with concomitant interferon therapy in L1210 tumor-bearing mice using relatively high doses (10^5 U/mouse/treatment) of the hybrid human interferon, IFN-αAD(Bgl).[66] This effect is less pronounced if the interferon treatments are delayed until after the cyclophosphamide treatments.[248] The negative interaction does not occur with melphalan in combination with IFN-αAD(Bgl). At lower doses of IFN-αAD (500 U/mouse/treatment) significantly greater antitumor effects occur with the combination treatments than with either interferon or cyclophosphamide alone and this effect also occurs with mouse interferon.[285] There is the possibility that patients on cyclophosphamide therapy may not obtain benefit from the cyclophosphamide during adventitious viral infections which induce interferon. Clearly, the possibility of negative interactions between cyclophosphamide or other drugs used with interferons needs to be borne in mind in monitoring clinical studies, particularly with high interferon doses.

As discussed above, IFN-αAD(Bgl) suppresses hepatic P-450 metabolism in the mouse. Although cyclophosphamide is converted to the active metabolite by the hepatic cytochrome P-450 system melphalan is unaffected by P-450 metabolism. Thus, superficially, there is a correlation between requirements for a drug to be metabolized by P-450, to be effective, and the negative interaction with interferon. However, in the rat, modulation of P-450 metabolism has been shown to have no effect on efficacy of cyclophosphamide, probably because drug dose × time is essentially unaltered.[211] It seems more likely that abrogation of the efficacy of cyclophosphamide arises from effects of the interferon on immune responses. The efficacy of IFN-αAD(Bgl) alone against the L1210 leukemia has been shown to be primarily by indirect immune mechanisms[162] and mouse interferons[212-214] as well as human leukocyte interferon[126,130] modulate immune responses. In addition to direct cytotoxic effects cyclophosphamide is known to stimulate some immune responses by inhibiting suppressor cell activity[215] and indirect effects on host antitumor immunity have been shown to be important for eradication of tumors by cyclophosphamide.[216,217] Suppression of tumor immunity by interferon could therefore account for abrogation of the efficacy of cyclophosphamide in combination therapies.

Other studies in mice, involving combinations of cyclophosphamide and interferons, have not shown negative interactions. Slater et al.,[218] using comparable doses of crude mouse interferon (2.8×10^5 U/treatment) against L1210 in a similar regimen to that of Lee et al.,[162] found no interaction with cyclophosphamide. O'Neill and Stebbing[219] found no interaction between *C. parvum* and cyclophosphamide and *C. parvum* is known to induce leukocyte and immune interferon production.[220,221] However, the interferon concentrations induced by *C. parvum* may be low and thus not sufficient to cause the negative interaction with cyclophosphamide. It is noteworthy that addition of specific antitumor antibodies enhanced the efficacy of combined cyclophosphamide and *C. parvum* therapy[219] as might be predicted from the studies[217] which indicate that antitumor immunity may be a limiting factor for cyclophosphamide therapy.

Some unexpected toxic effects of combination treatments are likely to occur. This seems to occur with combinations of Ara-A and natural IFN-α preparations in the treatment of hepatitis B patients.[222] Ara-A is not metabolized by the P-450 system and the cause of the toxic effects and whether they are peculiar to Hepatitis B patients remains obscure.

The effects of interferon on antibody synthesis are complex (see Section III.D.2) and the relation of in vitro studies on peripheral blood lymphocytes to in vivo effects is uncertain. The difference between data for human and mouse preparations has been related to the greater percentage of T cells in human peripheral blood.[130] Thus the pathological state, in particular T cell levels, may influence drug interactions in human clinical situations. The known immunosuppressive effects of interferons, combined with the apparent requirement for an immune response for effective tumor regression indicate that factors inhibiting suppressor T cells could augment the activity of interferons. Suppressor T cells are known to carry histamine 2 receptors which are blocked by cimetidine. The recent observation of dramatic improvement of melanoma patients on interferon therapy after addition of cimetidine[223] is thus encouraging for interventions that specifically modulate immune responses.

V. TOXICITY STUDIES AND SIDE EFFECTS

A number of symptoms observed after administration of interferon to humans have been considered side effects, including fever, chills, fatigue, joint pain, leukopenia, and anorexia. A listing of these side effects as reported in several studies involving at least nine patients is presented in Table 2. It is apparent that these effects occur with both crude and highly purified preparations of all types of human interferons, whether they are derived from natural sources, continuous cell lines, or genetically engineered bacteria. Fever, fatigue, and leukopenia are consistently the most commonly reported symptoms responsible for dose reduction or discontinuation of interferon therapy. However, all these abnormalities readily ameliorate within days after interferon administration is halted.

Table 2

OBSERVED SIDE EFFECTS OF INTERFERONS IN HUMAN CLINICAL TRIALS

Interferon	Evaluated	Number of Patients Exhibiting						Ref.
		Fever	Chills	Fatigue	Pain	Leukopenia	Anorexia	
Leukocyte (crude)	9	7	6	8	3	9	2	224
Leukocyte (crude)	11	11	—	11	6	11	8	225
Leukocyte (crude)	37	28	19	35	—	37	31	226
Leukocyte (crude)	16	16	8	2	—	13	—	177
Leukocyte (crude)	23	20	—	17	—	16	17	227
Leukocyte (crude)	29	18	11	3	11	—	—	228
Leukocyte (pure)	9	6	6	7	5	9	3	229
Fibroblast (crude)	26	24	22	10	—	5	1	230
Fibroblast (crude)	12	12	5	—	—	5	—	231
Bacterial IFN-αA (pure)	16	16	12	—	—	13	—	115, 177
Lymphoblastoid (pure)	9	9	—	6	—	5	9	232
Lymphoblastoid (pure)	42	41	28	42	2	42	42	184

Early studies showed that decreases in peripheral leukocytes, platelets, and reticulocytes were observed in chronic hepatitis patients treated with crude interferon preparations at doses as low as 8.5×10^4 U/kg/day.[233] In contrast, patients with varicella-zoster exhibited such effects only at interferon doses of 5.1×10^5 U/kg/day.[234] Polymorphonuclear leukocytes decreased within a day or two of initiation of treatment with a drop in reticulocytes occurring one week into therapy. Normal pretreatment levels of platelets, reticulocytes and polymorphonuclear leukocytes were reached 2 to 5 days after stopping interferon treatment.[234] These effects may be related to observations that in soft-agar assays growth of granulocyte-macrophage precursors from bone marrow, both of mice[235] and humans[205], are suppressed by interferon. Natural leukocyte interferon has been shown to arrest maturation of granulocyte precursors but it should be noted that not all patients showed abnormalities in differentiation of bone marrow cells grown in soft agar.[236]

The use of crude material in initial studies led to speculation about the factors responsible for the side effects.[237] The advent of sophisticated purification techniques and recombinant DNA technology has laid to rest the old argument of interferon vs. contaminants as the pyrogenic factor(s) in preparations. Clearly, highly purified homogeneous preparations of either mixtures of IFN-α subtypes,[229] a single IFN-α species,[115,177] or lymphoblastoid interferon[184] cause similar side effects. Comparative studies with semi-purified and highly purified natural IFN-α preparations obtained by monoclonal antibody affinity chromatography showed the same side effects as observed with crude material.[224,229] The first clinical trials with purified IFN-αA have shown that the symptoms produced by this interferon include fever, chills, myalgias, headaches, fatigue, and reversible leukopenia.[115,177] Somewhat unexpectedly, eight patients exhibited transient and mild numbness of the hands and/or feet.[177] As with trials utilizing crude natural material,[227] treatments with bacteria-derived interferon caused elevation of β_2-microglobulin levels.[226] Phase I studies with pure lymphoblastoid interferon have shown that doses as great as 100 to 200 $\times$ 10^6 U/m^2 can be tolerated but distinct neurological side effects accompany such high doses.[185] Several of the side effects noted for interferons may be acceptable, at least in the treatment of serious disease. However, the neurological effects seem, from recent unpublished results, to be the most serious and dose limiting problem associated with IFN-β and IFN-αs.

A potential problem associated with the administration of exogenous proteins for therapeutic purposes is the generation of antibodies to a normal human protein. Antibodies to human alpha interferon have been reported for a patient with systemic lupus erythematosus.[238] However, such a response dose not seem to be limited to individuals with this immunological disorder. Three of sixteen patients administered bacteria-derived IFN-αA developed antibodies to this interferon. These antibodies are apparently of the IgG class and were undetectable prior to the study and while the patients were receiving interferon.[115] Because in vitro neutralization of IFN activity occurs with these antibodies, one could speculate that in vivo neutralization is also possible. High antibody titers have been observed in two patients treated for many months with natural IFN-β and in one case the possibility of inability to produce IFN-β in the patient has been investigated.[239] Fibroblast cultures established from the patient were shown to be capable of producing IFN-β so that the development of antibodies is unlikely to be due to recognition of a foreign protein.[239] The patients' antibody titer against natural IFN-β was found to be four times greater than the titer against *E. coli* derived IFN-β_1,[286] indicating that the carbohydrate moiety of IFN-β may enhance the potential for production of auto-antibodies. Because there are several different IFN-α subtypes and it is unknown whether all individuals produce all subtypes it is possible that development of antibodies in patients treated with IFN-αs arises from recognition of determinants not present on those IFN-αs produced by the patient. However, there are other explanations for development of antibodies such as minor structural changes eliciting antibody responses due to lyophilization or the method of delivery. To date, no neutralizing antibody titers have been detected in patients receiving lymphoblastoid interferon.[287] The ramifications of these observations in the clinical use of highly purified interferon preparations will have to await further clinical studies.

The use of animal models to predict the side effects mentioned, particularly elevation of body temperature and leukopenia, has met with limited success. The administration of IFN-αA, -αD, and -αAD(Bgl) to squirrel monkeys did not result in a significant drop in white blood cell counts. However, the injection or these IFN-α subtypes or IFN-β_1 from *E. coli* into rhesus monkeys did cause a marked suppression in peripheral leukocyte numbers.[288] In contrast, the administration of bacteria-derived IFN-α_2 did not result in suppressed leukocyte numbers or elevated body temperatures in rhesus monkeys, whereas natural material from buffy coat cells did.[149] Recently, Finter and co-workers have reported no significant side effects when highly purified lymphoblastoid interferon is given to either patas or rhesus monkeys.[240] Although highly purified human interferon is not innocuous in man, it is apparently rather well tolerated by sub-human primates at relatively high doses. However, the outcome of long-term administrations of interferons to humans could be rather questionable.

Studies performed in murine systems with homologous interferon have revealed some rather deleterious effects. The administration of interferon to newborn mice can cause death at high doses[241] and causes a marked wasting effect in these animals when given over a prolonged period of time.[242] These effects could be related to the anticellular properties of the interferon or the cellular toxicity of the material as reported for the hepatic systems. These animal models have shown that interferon injections also cause glomerulonephritis in mice[243] and inhibit the regeneration of liver following partial hepatectomy.[244] Murine interferons have been shown to affect immune responses both in vivo and in vitro and cause alterations in lymphoid cell function as well as thymocyte surface markers.[245] Interferons may also affect distribution of cells in vivo. For example, lymphoid suppression by cortisone treatment apparently is due to limitation of release of cells from bone marrow[246] and determination of bone marrow cellularity would appear to be important to quantitate these effects. Although some of the early studies in murine systems were performed with crude interferon, the more recent work has involved purified preparations. More importantly, recent experiments employing a homogeneous preparation of the hybrid human leukocyte interferon IFN-αAD, which has antiviral activity in mice, have shown that inhibition of the hepatic P-450 system occurs.[144] It is unknown whether this effect is related to the toxicity observed in newborn mice. Such studies in animal systems dictate that interferons cannot be used indiscriminately in man but must be subjected to careful clinical evaluations.

VI. STRUCTURE/ACTIVITY RELATIONS

Although studies reviewed here indicate interferon functions which may be separable there are as yet no definitive data relating particular activities with particular structural domains within the interferons. The generation of hybrid IFN-αs has allowed initial forays into interferon structure/function studies but has not provided any real clues to structural domains involved in specific interferon functions. In addition to the IFN-αA and -αD hybrids, cited in Sections II, III, and IV, a series of hybrids involving a common *Pvu*II restriction site has been formed using IFN-αA, -αD and other IFN-α subtypes: IFN-αB, -αF, and -αG.[248] The majority of these interferons have no activity on mouse L-929 cells while all have significant activity on the bovine MDBK cells. Only one interferon, the IFN-αFD hybrid, appears to have approximately equal activity on human, mouse, bovine, and monkey kidney cell lines. Interestingly, the only other hybrids having activity on murine cells contain the C-terminus of IFN-αF (viz. IFN-αAF and -αBF) but these have markedly lower activities on MDBK cells. Whether the IFN-αFD hybrid interferon is active in vivo as described for IFN-αAD(Bgl) is presently unknown.

The presence of a *Bgl*II or *Sau*3A restriction site at a position corresponding to residue 150 in IFN-αA, -αD and -αI also has allowed construction of other hybrids. These hybrids contain 165 or 166 amino acid residues, depending on whether or not they contain the N-terminal portion of IFN-αA, and they allow examination of the effect of modifying the C-terminal 15 amino acid residues. The IFN-αAD(Bgl/Sau) hybrid which includes only the 15 C-terminal residues of IFN-αD shows antiviral activity comparable to that of IFN-αAD(Bgl). The IFN-αIA(Bgl/Sau) and IFN-αDA(Bgl/Sau) hybrids have low activity. Thus modifications of the C-terminal 15 amino acids clearly have dramatic effects on the activities of these interferons. However, limited proteolytic cleavage of IFN-αA, resulting in loss of the C-terminal 13 residues, results in a material retaining nearly all antiviral, cell binding, and antibody binding activity.[12] Moreover, the human cell derived interferons lacking the 10 C-terminal amino acids found on their cloned counterparts[38] have biological activity. The *Bgl*II/*Sau*3A site at residue 150 of IFN-αA has also been used to construct a gene which codes for a derivative interferon (IFN-αA-11) lacking the last 11 amino acid residues. Curiously, this interferon has greatly reduced antiviral activity.[249] Presumably folding of IFN-αA-11 during synthesis results in a structure distinct from that obtained by post-translational cleavage of the C-terminus.

Although the C-terminal 13 residues of IFN-αA appear to be unnecessary for antiviral activity, at least when removed after synthesis, the majority of the rest of the molecule seems to be essential. Neither trypsin and cyanogen bromide fragments nor reconstituted mixtures of fragments have antiviral or cell binding activity.[12] The *cys* 29 to *cys* 138 disulphide bond of IFN-αs appears to be essential for activity because limited oxidation products are only active if this bond is maintained.[12] That the other disulphide, *cys* 1 to *cys* 98 is not essential was implied by the activity of fusion proteins lacking *cys* 1.[6] A fusion protein containing IFN-α_2 with 19 extra N-terminal residues from both the signal sequence and β-galactosidase is as active as IFN-α_2[250] indicating that addition of an N-terminal sequence does not affect antiviral activity. At least one of the two tryptophans in IFN-αA appears to be important for activity: chemical modification of the tryptophans greatly reduced antiviral, antigenic, and cell receptor binding activity.[12]

The formation of hybrids with the same N- and C-termini but different internal sequences allows analysis of the functional role of other residues. The various IFN-αAD hybrids illustrate this approach and most extensive pharmacological data exist for IFN-αAD(Bgl) and IFN-αAD(Pvu). The only differences between these two materials are in the amino acid sequences of IFN-αA and -αD between residues 61 and 91 and these are at positions 68, 79, and 85.[61] These residues are thr, asp, cys in IFN-αAD(Bgl) and ser, thr, tyr in IFN-αAD(Pvu).

The specific activities of IFN-αA, -αD, -αAD(Bgl), and -αAD(Pvu) are essentially identical when determined in bovine MDBK cultures challenged with VS virus. In sharp contrast, their specific activities vary greatly when measured on other cell types challenged with either EMC or VS virus. In mouse cells IFN-αAD(Bgl) and -αAD(Pvu) show a difference in specific activity of about fourfold or tenfold depending on the virus challenge.[64] In vivo studies have also revealed significant differences between the biological activities of these two hybrids. IFN-αAD(Bgl) appears to be much more potent than IFN-αAD(Pvu) against L1210 leukemia in mice but the dose response curves in this tumor model are quite different. Above 10^4 U/treatment IFN-αAD(Bgl) shows decreasing efficacy whereas the efficacy of IFN-αAD(Pvu) decreases only at doses over 10^5 U/treatment.[61] These results are much different from those generated in an antiviral mouse model where only IFN-αAD(Bgl) showed significant protection against lethal doses of EMC virus. Dose response studies have demonstrated that antiviral protection plateaus at high doses but there is no decrease in efficacy.[65] In contrast, both hybrid interferons are active in hamsters against lethal doses of EMC virus and IFN-αAD(Pvu) treated animals had slightly greater survival times.[288] Thus, these two unique hybrid interferons not only differ in their actions on tumor cells and virus growth in animals but also in their abilities to cross species lines.

As described above (Section III.E) IFN-αAD(Bgl) causes suppression of mouse P-450 metabolism[144] with predictable effect on the activity of hexabarbital and acetaminophen, both of which are metabolized by P-450.[145] However, IFN-αAD(Pvu) does not affect metabolism of hexabarbital or acetaminophen.[61] The pharmacokinetics in rats of these two hybrid interferons are also different. Although nephrectomization decreases clearance of plasma interferon concentration this effect is much more prominent for IFN-αAD(Pvu) than IFN-αAD(Bgl). All these differences arise from the three amino acid residue differences between these materials with very similar primary structure.

Studies in squirrel monkeys infected with EMC virus have allowed comparison of IFN-αA, -αD, and -αAD(Bgl) and have shown that the hybrid interferon is at least as effective as the parental molecules in this model (See Section III.F). While IFN-αA causes leukopenia in rhesus monkeys, this does not occur in squirrel monkeys.[251] However, efficacy against viral infections has been demonstrated with this interferon subtype in both primate species[149,150] indicating that leukopenia is not intrinsically linked with antiviral effects.

VII. CLINICAL IMPLICATIONS

Potential clinical interest in interferons has focused on viral infections and neoplastic diseases. Antiviral applications were inherently obvious from the assay involved in the discovery of the interferons and indirect cytostatic and cytotoxic mechanisms have provided a basis for anticipating antitumor effects. The number of patients who have been treated clinically with interferons remains relatively small and the most susceptible viral infections and tumors are still not apparent. This arises from the observations that, in vivo, indirect effects not observed in antiviral or antiproliferative cell culture assays may be of primary importance and the uncertainty as to which assays, if any, are predictive of efficacy. Although limited, clinical data are unusual for the current stage of development of interferons as drugs and careful consideration of known effects should greatly assist their application in clinical medicine. Identification of potential clinical indications is facilitated by current clinical and preclinical data together with analysis of the mechanisms of action of interferons that appear to be of importance and the possibilities and properties of novel recombinant DNA derived interferons.

A. Data from Clinical Trials: Potential Indications

The interpretation of the body of data generated to date in clinical antitumor trials is somewhat difficult. Interpretation of data is hampered by the fact that most reports describe nonrandomized pilot studies with material of low purity. The most extensive antitumor trials conducted to date have been in patients with osteogenic sarcoma. Although the study of Stander et al.[252] was not a truly randomized one, the results after 5 years appear encouraging. As compared to a contemporary group of untreated individuals, the interferon treated group had less disease and fewer deaths.[252] Leukocyte interferon has also been used in studies of multiple myeloma, malignant lymphoma, and various leukemias. However, because of the small groups of patients entered in studies at a number of different centers, interpretation of results is difficult but overall the incidence of positive responses is rather low.

Patients with non-Hodgkins lymphoma also have been treated with interferon. In two separate studies the best responses were seen in individuals with the nodular, poorly differentiated lymphocytic form of this disease.[253] Regression of lymph node masses occurred in three patients within 2 weeks of beginning interferon therapy. Following treatment, remission of disease continued for 6 to 12 months in the absence of additional interferon treatment. However, only about $1/4$ of the total patient population showed this type of response to the interferon injections.

One of the few solid tumors to be treated with leukocyte interferon is breast carcinoma. Most treatment schedules have utilized intramuscular administration of between 3 and 9 million units of interferon daily for a period of 4 weeks. A recent study involving twenty-three patients and using buffy coat interferon demonstrated partial responses in five patients.[227] The responders were significantly older than the rest of the group and all individuals presented the side effects normally observed with interferon therapy. Experience with other solid tumors has included studies with malignant melanoma and non-small cell lung carcinoma patients. No complete remissions were observed in these studies.[254,255]

The potential for clinical success with interferons would seem to be greatest for diseases of viral etiology or those where virus infection is implicated. There is now strong evidence for the involvement of human papilloma viruses in juvenile laryngeal papillomatosis and this condition has shown a positive response to interferon treatment.[256] Virus infections are quite prevalent in bone marrow transplant patients who routinely suffer from herpes simplex, herpes zoster, and cytomegalovirus outbreaks. A study with seven of these patients showed that although leukocyte interferon did not reduce mortality, the mean titer of cytomegalovirus was reduced in five patients.[257] Perhaps the prophylactic treatment of such individuals would result in a significant reduction in virus infection without affecting bone marrow cell survival. Interferon therapy may prove valuable for virus infections as mild as the common cold as well as in life-threatening situations. Clinical trials conducted with either purified natural[258] or bacteria-derived[259] IFN-α have shown a reduction in the incidence and severity of colds. Mean clinical scores, nasal secretions, and virus secretion were all significantly reduced in interferon-treated patients.[259] In a study of 15 critically ill patients with life-threatening infections, 12 recovered after the administration of natural leukocyte interferon.[260] Other viral infections with clinical evidence for efficacy of natural interferon preparations include hepatitis B,[261,262] herpes simplex[263] and zoster,[264] and adenovirus keratoconjunctivitis.[265]

Interferon inducers have been shown to inhibit ocular lesions due to *Chlamydia trachomatis*[265] and interferons have been shown to inhibit cell replication of chlamydia.[267] An increase in interferon titers has been observed in children infected with *Plasmodium falciparum* and this was positively correlated with NK cell activity as well as the degree of parasitemia.[268] However, the NK cell activities were lower than expected for the interferon concentrations observed and enhanced NK cell activity could not be achieved on incubation of purified blood lymphocytes with interferon. These effects may be specific to children or the parasite species because no circulating interferon has been observed in adults infected with *Plasmodium vivax*.[269] Deficiency in NK cell activity in Chediak-Higashi patients[270] indicates that such patients may benefit from interferon therapy during the chronic immunosuppressed phase of the disease or during the terminal lymphoproliferative disorder. Other clinical situations in which interferons could be useful include leishmaniasis,[271] ulcers, and deep burns.[272]

Now that many of the problems surrounding purification and availability of interferons are being overcome, well-designed clinical studies should lead to definitive identification of the most susceptible diseases. However, taking together all the current preclinical and clinical data there is no doubt that a range of human viral infections respond very dramatically to interferon therapy. The data for human tumors are not so promising for IFN-α and IFN-β, but progress in understanding relevant mechanisms of action and empirical optimization of treatment regimens could lead to significant clinical progress. It seems regrettable that so little clinical experience has been obtained for use of interferons in viral infections, because such data and treatment regimens and doses found to be effective could influence and benefit applications to other diseases, including neoplasias.

B. Mechanisms of Action of Interferons

Investigations of the mechanisms of action of human interferons are of obvious importance for clinical development of these materials. Some studies can be readily carried out in human cell cultures and periperal blood, but the limitations of such studies is becoming obvious. Use of the hybrid human interferons in mice has allowed investigation of mechanisms of action and some principles relevant for effective treatment regimens. Although the relevance to human clinical diseases of murine models, particularly transplantable tumors, may be questionable, some conclusions drawn from such studies may be applied to human clinical trials and their relevance thereby tested. The apparently fortuitous activity of the hybrid interferon IFN-αAD(Bgl) in mice has allowed study of biological effects of relevance to human clinical use. It is perhaps remarkable that the hybrid interferon IFN-αAD(Bgl) with pronounced activity in mouse cells has also shown antiviral and antitumor activity in mice comparable to mouse interferons. Furthermore, effects on P-450 metabolism and in vivo drug interactions appear to be very similar for IFN-αAD(Bgl) and mouse interferons. It seems unlikely that this arises from coincidental construction of an interferon whose entire amino acid sequence is closely similar to a mouse interferon. Studies with human interferons and their analogs in heterologous systems is a necessity but may also provide information that could not be readily obtained even in primate studies. This is further illustated by the effects of IFN-αA on the efficacy of cyclophosphamide against the TBD932 lymphosarcoma of hamsters, in which species the interferon alone has no significant antiviral or antitumor activity. The purpose of such studies in heterologous systems should be distinguished from studies designed to define protocols of direct relevance to clinical use. For the latter purpose primate studies are likely to remain most significant. Although some useful primate models exist for viral infections, they have not been widely explored and useful tumor models are still very limited.

Understanding the relation of cell culture to in vivo effects of interferons remains a problem of considerable practical as well as theoretical interest. The primary mechanism of action of some chemotherapeutic agents may not be due to the direct antiproliferative effects observed in cell cultures. However, correlations have been observed between cell culture effects and efficacy and features such as pharmacokinetics have often proved useful in developing optimum treatment regimens.

Cell culture assays are obviously useful to the extent that they reflect mechanisms of action that operate in vivo. However, the mechanisms of importance may be different for different diseases. The issue under consideration is thus central to understanding mechanisms of action of interferons and the relation of antiviral to antitumor effects. Too often activities of interferons in cell assays have been assumed to relate to in vivo efficacy. Indeed, identification of mechanisms of importance for efficacy from appropriate animal studies could more usefully precede studies of in vitro effects of uncertain significance. The reverse has generally occurred, resulting in analysis of mechanisms which may have little clinical significance. In every case it is also necessary to estimate the quantitative role of any supposed mechanism of action to overall antiviral and antitumor effects. Although any particular mechanism, such as NK-cell mediated cell lysis may play a role in interferon therapy of defined diseases, the mechanisms may be of primary importance for some diseases and of only minor significance in other diseases. Some proposed mechanisms may be of no quantitative significance in certain diseases.

It is now recognized that interferons and other agents which modify indirect mechanisms affecting disease may require new approaches to preclinical and clinical studies if full clinical potential is to be realized. At least for antitumor agents, direct antiproliferative effects may not be as important as inhibitory or other effects on particular cells which then cause tumor regression. Direct antiviral and antiproliferative effects of interferons in cell cultures did identify these agents as of potential clinical utility for viral infections and neoplastic diseases. These and the other biological effects of interferons need to be borne in mind when designing clinical studies. However, attention to other possible modes of action is likely to prove critical for successful use of interferons and derivative agents in clinical medicine. Conventional chemotherapeutic agents have generally been developed when the direct mechanism of action has been understood, or thought to be understood, and such features as blood concentrations, in particular half-life and the minimal effective dose have been assessed. However, features such as those related to blood concentration may prove irrelevant for natural agents for which there are cellular receptors of high affinity but low density on cell surfaces. More critical, however, are assumptions regarding the mechanism of action based on the various biological effects observed in bioassays. There is a need to recognize that early clinical studies should be designed to test whether particular mechanisms of action, identified in preclinical studies, are of importance.

Current recognition that interferons have a range of different biological effects should alert us to the fact that natural physiological substances generally show quanitatively different effects over different dose ranges. This is a feature of peptide hormones in general which, at pharmacological doses, show effects not observed at physiological doses.[273] Moreover, the transient high doses achieved after bolus administration may have long lasting effects that do not occur after regulated administration. These factors are highly pertinent to the interferons and should be carefully considered. Real progress towards clinical use of interferons is now likely to occur only by investigating mechanisms of action in relation to the dose dependency of different effects.

C. Novel Interferons by Recombinant DNA Methods

As reviewed above, the construction of hybrid interferons has provided an initial method for probing structure activity relationships of interferons. Most current data relate to IFN-αs because they are a family of related genes with common restriction enzyme cleavage sites. Additional structurally related genes have not yet been identified and expressed in microorganisms by recombinant DNA methods for IFN-β_1 and IFN-γ_1. However, new analogs may be most conveniently produced by complete synthesis of genes[274] rather than elaborate construction of hybrid genes and this approach will be particularly applicable for constructing analogs of IFN-β_1 and IFN-γ_1. It should be noted that the initial analogs may only indicate amino acid positions or peptide regions associated with particular properties. For the different IFN-αs the distinctive properties of particular subtypes must be associated with at least some of the residues peculiar to the subtype in question. Thus for IFN-αs there is some guide to sequences which confer subtype specific effects. However, at present these data are most useful in indicating regions which are unlikely to be involved in specific effects because these should include the extensive sequences common to all the IFN-α subtypes (see Figure 1).

Study of the properties of the initial analogs, however they are produced, provides empirical data relating structure to activity. It is apparent that alteration of particular amino acid residues can have profound effects on the properties of particular interferons. However, it is not clear how many other amino acid changes could produce the same effect and whether other changes would affect a different spectrum of biological properties. The empirical phase of probing structure/activity relationships of interferons has only just begun. This phase should lead to a deductive phase in which further analogs may be designed with particular desired properties. Their properties may be a subset of those of the original materials or a combination of properties considered to be suitable for treatment of particular diseases. Current data for the hybrid interferons have extended the principles of structure activity studies to macromolecules, but adequate information for predicting the properties of particular novel interferon-like sequences has not yet accumulated.

D. Factors Determining Clinical Development

The varied biological effects of IFNs, differences in the properties of molecular species obtained by cloning individual genes in microorganisms, and differences in the response of target cells compound the difficulties encountered in designing and interpreting clinical studies. Various factors contributing to clinical efficacy, determined from preclinical studies and early clinical results, have been reviewed here. For clinical efficacy these factors may be considered as one relates primarily to the molecular species of the interferon or the nature of the disease in the patient.

Although absence of glycosylation of IFN-β and IFN-γ produced in *E. coli* has not yet been observed to affect any of the biological activities examined, it remains possible that important differences from the natural materials will be found. Comparison of natural and microbial materials remains of great importance. Investigation of the properties of mixtures of purified materials could also be informative, particularly different compositions of the various IFN-α subtypes. It seems likely that there are interactive effects between interferons and other lymphokines and also between IFN-γ and the other classes of interferons. It is also important to bear in mind that even highly purified IFNs produced from microorganisms may contain trace amounts of bacterial contaminants which could affect the biological activity of the interferons.

There can be no doubt that understanding the primary mechanisms of action elicited by an interferon are likely to be the most important factors in clinical development of that interferon or derivative agents for treatment of particular diseases. Detailed studies of interferon doses and their effects on a range of physiological parameters is likely to prove informative. Although plasma concentrations of interferons may not be directly related to efficacy, the interferon dose administered can determine the nature and magnitude of effects, as described above for many in vitro and in vivo effects. Knowledge of which effects are stimulated in vivo at different interferon doses could be particularly useful for development of clinical regimens and uses. Natural biological agents generally have multiple effects with different effects occurring at different doses. This is the case even for an agent such as insulin, whose clinical use is aimed primarily at one effect.[273]

In murine systems there are genetic factors which influence the efficacy of interferons. The *Mx* locus affects the efficacy of interferons specifically against influenza virus infections without affecting efficacy against other viruses such as EMC virus.[275] Moreover, balb/c mice are more sensitive to the effect of interferon against EMC virus that C57Bl/6 mice.[276] Analysis of back-crosses indicates dominance for greater sensitivity to interferon but also the existence of multifactorial genetic factors. However, fibroblasts derived from the two strains of mice are equally sensitive to the antiviral effects of the interferon, indicating an indirect mechanism for antiviral activity.[276] Cells from the hemopoietic system of the two mouse strains are differentially sensitive to antiproliferative effects of the interferon, whereas embryo fibroblasts are not,[276] indicating that not all cells express the differential interferon sensitivity and the effect may not be manifest in both antiviral and antiproliferative effects. No doubt comparable genetic factors modulate the efficacy of interferons in the human population. However, the only known locus is one on chromosome 21 which causes increased sensitivity to interferon in trisomic cells from patients with Down's syndrome.[277]

In the case of virus infections that induce interferons there may be modifying host responses. Comparison of the infectivity of wild-type and an interferon-sensitive strain of Mengo virus in mice shows that both are equally infectious at low doses but that at high doses the interferon sensitive strain of virus causes decreased incidence of deaths.[278] Similarly, HSV-1 causes a lethal infection of mice at low virus doses but higher doses are not lethal because they cause induction of interferon.[279] The interferon-sensitive strain of Mengo virus was sensitive to IFN-β but not to IFN-α.[278] Because IFN-β is the form of interferon induced naturally by infection with Mengo virus, it is apparent that selection of altered sensitivity to this infection may occur naturally. For this virus there is no obvious selection pressure for altered sensitivity to IFN-α, which also causes more pronounced antiviral effects than IFN-β even with the sensitive virus strain.[278] Such observations are important insofar as they indicate whether resistance to interferon therapy may occur and to which interferon resistance may develop.

A factor of importance, particularly in the treatment of neoplasia, is whether increasing doses of an interferon can achieve significant clinical efficacy. The involvement of many mechanisms of action and other factors makes it possible that non-interferon mediated factors become limiting, at least in some diseases. For mechanisms involving specific antibodies such as antibody-dependent cell mediated cytotoxicity, antibody concentration or macrophage Fc receptors could limit efficacy. Antibodies that inhibit tumor growth appear to require Fc receptors[280] and interferons may be beneficial in some cases only insofar as they stimulate Fc receptor synthesis. Passive serotherapy is also known to augment the antitumor efficacy of chemotherapeutic and immuno-stimulating factors.[219] Understanding the balance of interferon and non-interferon mechanisms and their interaction provides the major challenge for future development of interferons as useful clinical agents.

ACKNOWLEDGMENT

The authors are indebted to Linda Fenton for her skill and patience in typing and retyping the various drafts of this text.

REFERENCES

1. **Baron, S., Buckler, C. E., Friedman, R. M., and McKloskey, R. V.,** Role of interferon during viremia. I. Production of circulating interferon, *J. Immunol.,* 96, 12, 1966.
2. **Finter, N. B.,** Interferon as antiviral agent *in vivo:* quantitative and temporal aspects of potection of mice against Semliki Forest Virus, *Br. J. Exp. Pathol.,* 47, 261, 1966.
3. **Pauker, K., Cantell, K., and Henle, W.,** Quantitative studies on viral interference in suspended L-cells. III. Effect of interfering viruses and interferon on the growth rate of cells, *Virology,* 17, 324, 1962.
4. **Zoon, K. C., and Wetzel, R.,** Comparative structures of mammalian interferons, in *Handbook of Experimental Pharmacology,* Came, P. E. and Carter, W. A., Eds., Springer-Verlag, New York, 1983, 71.
5. **O'Malley, J. A., Nussbaum-Blumenson, A., Sheedy, D., and Grossmayer, B. J.,** Identification of the T cell subset that produce human gamma interferon, *J. Immunol.,* 128, 2522, 1982.
6. **Nagata, S., Taira, H., Hall, A., Johnrud, L., Streuli, M., Ecsodi, J., Boll, W., Cantell, K., and Weissmann, C.,** Synthesis in *E. coli* of a polypeptide with human leukocyte interferon activity, *Nature (London),* 284, 316, 1980.
7. **Nagata, S., Mantes, N., and Weissmann, C.,** The structure of one of the eight or more distinct chromosomal genes for human interferon -α, *Nature (London),* 287, 401, 1980.
8. **Goeddel, D. V., Leung, D. W., Dull, T. J., Gross, M., Lawn, R. M., McCandiss, R., Seeburg, P. H., Ullrich, A., Yelverton, E., and Gray, P. W.,** The structure of eight distinct cloned human leukocyte interferon cDNAs., *Nature (London),* 290, 20, 1981.
9. **Lawn, R. M., Adelman, J., Dull, T. J., Gross, M., Goeddel, D. V., and Ullrich, A.,** DNA sequence of two closely linked human leukocyte interferon genes, *Science,* 212, 1159, 1981.
10. **Brack, C., Nagata, S., Mantei, N., and Weissmann, C.,** Molecular analysis of the human interferon-alpha gene family, *Gene,* 15, 379, 1981.
11. **Wetzel, R., Perry, L. J., Estell, D. A., Lin, N., Levine, H. L., Slinker, B., Fields, F., Ross, M. J., and Shively, J.,** Properties of a human alpha-interferon purified from *E. coli* extracts, *J. Interferon Res.,* 1, 381, 1981.
12. **Wetzel, R., Levine, H. L., Estell, D. A., Shire, S., Finer-Moore, J., Stroud, R. M., and Bewley, T. A.,** Structure-function studies on human alpha interferon, in *Chemistry and Biology of Interferons,* Vol. 25, Merigan T., Friedman, R., and Fox, C. F., Eds., Academic Press, New York, 1983, 365.
13. **Rubinstein, M., Levy, W. P., Moschera, J., Lai, C. Y., Hershberg, R. D., Bartlett, R. T., and Pestka, S.,** Human leukocyte interferon: isolation and characterization of several molecular forms, *Arch. Biochem. Biophys.,* 210, 307, 1981.
14. **Allen, G. and Fantes, K. H.,** A family of structural genes for human lymphoblastoid (leukocyte-type) interferon, *Nature (London),* 287, 408, 1980.
15. **Berg, K. and Heron, I.,** Human leukocyte interferon comprises a continuum of 13 interferon species, in *Human Lymphokines: Biological Response Modifiers,* Khan, A. and Hill, N. O., Eds., Academic Press, New York, 1982, 397.
16. **Tanaguchi, T., Ohno, S., Fujii-Kuriyama, Y., and Muramatsu, M.,** The nucleotide sequence of human fibroblast interferon cDNA, *Gene,* 10, 11, 1980.
17. **Derynck, R., Conent, J., DeClercq, E., Volchaert, G., Tavernier, J., Devos, J., and Fiers, W.,** Isolation and structure of a human fibroblast interferon gene, *Nature (London),* 285, 542, 1980.
18. **Goeddel, D. V., Shepard, H. M., Yelverton, E., Leung, D., and Crea, R.,** Synthesis of human fibroblast interferon by *E. coli, Nucleic Acids Res.,* 8, 4057, 1980.
19. **Houghton, M., Eaton, M. A. W., Stewart, A. G., Smith, J. C., Doel, S. M., Catlin, G. H., Lewis, H. M., Patel, T. P., Emtage, J. S., Carey, N. H., and Porter, A. G.,** The complete amino acid sequence of human fibroblast interferon as deduced using synthetic oligodeoxyribonucleotide primers of reverse transcriptase, *Nucleic Acids Res.,* 8, 2885, 1980.
20. **Pitha, P. M., Slate, D. L., Raj, N. B. K., and Ruddle, F. H.,** Human β interferon gene localization and expression in somatic cell hybrids, *Mol. Cell Biol.,* 2, 564, 1982.
21. **Weissenbach, J., Chernajovsky, Y., Zeevi, M., Shulman, L., Soveq, H., Niv, U., Wallach, D., Perncaudet, M., Trollarv, P., and Revel, M.,** Two interferon mRNAs in human fibroblasts: *in vitro* translation and *Escherichia coli* cloning studies, *Proc. Natl. Acad. Sci. U.S.A.,* 77, 7152, 1980.
22. **Sehgal, P. B. and Sagar, A. D.,** Heterogeneity of poly(I).poly(C)-induced human fibroblast interferon mRNA species, *Nature (London),* 238, 95, 1980.
23. **Gray, P. W., Leung, D. W., Pennica, D., Yelverton, E., Najarian, R., Simonsen, S. S., Derynck, R., Sherwood, P. J., Wallace, D. M., Berger, S. L., Levinson, A. D., and Goeddel, D. V.,** Expression of human immune interferon cDNA in *E. coli* and monkey cells, *Nature (London),* 295, 503, 1982.
24. **Devos, R., Cheroutre, H., Toya, Y., Degrave, W., Van Heuverswyn, H., and Fiers, W.,** Molecular cloning of human immune interferon cDNA and its expression in eukaryotic cells, *Nucleic Acids Res.,* 10, 2487, 1982.

25. **Goldstein, L. D., Langerford, M. P., Stanton, G. J., Deley, M., and Georgiades, J. A.,** Human gamma interferon: different molecular species, in *The Biology of the Interferon System*, DeMaeyer, E., Galasso, G., and Schellekens, H., Eds., Elsevier/North Holland, Amsterdam, 1981, 313.

26. **Havell, E. A., Yip, Y. K., and Vilcek, J.,** Characteristics of human lymphoblastoid (Namalva) interferon, *J. Gen. Virol.*, 38, 51, 1977.

27. **Dalton, B. J. and Paucker, K.,** Antigenic properties of human lymphoblastoid interferons, *Infect. Immun.*, 23, 244, 1979.

28. **Traub, A., Feinstein, S., Grosfeld, H., Payess, B., Lazar, A., Reuveny, S., and Mizrahi, A.,** Isolation of two species of human lymphoblastoid (Namalva) cell-derived interferon that differ in rates of clearance, size and cell specificity, *J. Interferon Res.*, 1, 571, 1981.

29. **Morser, J., Meager, A., Burke, D. C., and Secher, D. S.,** Production and screening of cell hybrids producing a monoclonal antibody to human interferon alpha, *J. Gen. Virol.*, 53, 257, 1981.

30. **Allen, G., Fantes, K. H., Burke, D. C., and Morser, J.,** Analysis and purification of human lymphoblastoid (Namalva) interferon using a monoclonal antibody, *J. Gen. Virol.*, 63, 207, 1982.

31. **Dworkin-Rastl, E., Dworkin, M. B., and Swetly, P.,** Molecular cloning of human alpha and beta interferon genes from namalva cells, *J. Interferon Res.*, 2, 575, 1982.

32. **Evinger, M., Rubinstein, M., and Pestka, S.,** Antiproliferative and antiviral activities of human leukocyte interferons, *Arch. Biochem. Biophys.*, 210, 319, 1981.

33. **Imai, M., Sano, T., Yanase, Y., Mayamoto, K., Yonehara, S., Mori H., Honda, T., Fukuda, S., Nakamura, T., Miyakawa, Y., and Mayumi, M.,** Demonstration of two subtypes of human leukocyte interferon (IFN-α) by monoclonal antibodies, *J. Immunol.*, 128, 2824, 1982.

34. **Meurs, E., Rougeot, C., Svab, J., Laurent, A. G., Hovanessian, A. G., Robert, N., Gruest, J., Montagnier, L., and Dray, F.,** Use of an anti-human leukocyte interferon monoclonal antibody for the purification and radioimmunoassay of human alpha interferon, *Infect. Immun.*, 37, 919, 1982.

35. **Weissmann, C.,** The cloning of interferon and other mistakes, in *Interferon*, Vol. 3, Gresser, I., Ed., Academic Press, London, 1981, 101.

36. **Zoon, K. C., Smith, M. E., Bridgen, P. J., Anfinsen, C. B., Hunkapiller, M. W., and Hood, L. E.,** Amino terminal sequence of the major component of human lymphoblastoid interferon, *Science*, 207, 527, 1980.

37. **Levy, W. P., Shively, J., Rubinstein, M., Valle, U. D., and Pestka, S.,** Aminoterminal amino acid sequence of human leukocyte interferon, *Proc. Natl. Acad. Sci. U.S.A.*, 77, 5102, 1980.

38. **Levy, W. P., Rubinstein, M., Shively, J., Valle, U. D., Lai, C. Y., Moschera, J., Brink, L., Gerber, L., Stein, S., and Pestka, S.,** Amino acid sequence of a human leukocyte interferon, *Proc. Natl. Acad. Sci. U.S.A.*, 78, 6186, 1981.

39. **Lawn, R. M., Gross, M., Houck, C. M., Franke, A. E., Gray, P. W., and Goeddel, D. V.,** DNA sequence of a major leukocyte interferon gene, *Proc. Natl. Acad. Sci. U.S.A.*, 78, 5435, 1981.

40. **Owerbach, D., Rutter, W. J., Shows, T. B., Gray, P. W., Goeddel, D. V., and Lawn, R. M.,** Leukocyte and fibroblast interferon genes are located on human chromosome 9, *Proc. Natl. Acad. Sci. U.S.A.*, 78, 3123, 1981.

41. **Slate, D. L., D'Eustachio, P., Pravtcheva, D., Cunningham, A. C., Nagata, S., Weissmann, C., and Ruddle, F. H.,** Chromosomal location of a human alpha interferon gene family, *J. Exp. Med.*, 155, 1019, 1982.

42. **Lawn, R. M., Adelman, J., Franke, A. E., Houck, C. M., Gross, M., Najarian, R., and Goeddel, D. V.,** Human fibroblast interferon gene lacks introns, *Nucleic Acids Res.*, 9, 1045, 1981.

43. **Houghton, M., Jackson, I. J., Porter, A. G., Doel, S. M., Catlin, G. H., Barber, C., and Carey, N. H.,** The absence of introns within a human fibroblast interferon gene, *Nucleic Acids Res.*, 9, 247, 1981.

44. **Tavernier, J., Derynck, R., and Fiers, W.,** Evidence for a unique human fibroblast interferon (IFN-beta 1) chromosomal gene, devoid of intervening sequences, *Nucleic Acids Res.*, 9, 461, 1981.

45. **Nagata, S., Brack, C., Henco, K., Schambock, A., and Weissman, C.,** Partial mapping of ten genes of the human interferon-alpha family, *J. Interferon Res.*, 1, 333, 1981.

46. **Ullrich, A., Gray, P., Goeddel, D. V., and Dull, T. J.,** Nucleotide sequence of a portion of human chromosome 9 containing a leukocyte interferon gene cluster, *J. Mol. Biol.*, 156, 467, 1982.

47. **Shows, T. B., Sakaguchi, A. Y., Naylor, S. L., Goeddel, D. V., and Lawn, R. M.,** Clustering of leukocyte and fibroblast interferon genes on human chromosome 9, *Science*, 218, 373, 1982.

48. **Proudfoot, N. J., Shander, M. H. M., Manley, J. L., Geffer, M. L., and Maniatis, T.,** Structure and *in vitro* transcription of human globin genes, *Science*, 209, 1329, 1980.

49. **Gray, P. W., and Goeddel, D. V.,** Structure of the human immune interferon gene, *Nature (London)*, 298, 859, 1982.

50. **Linsley, P. S. and Siminovitch, L.,** Comparison of phenotypic expression with genotypic transformation by using cloned, selectable markers, *Mol. Cell. Biol.*, 2, 593, 1982.

51. **Harkins, R. N., Hass, P. E., Kohr, W. H., Aggarwal, B. B., Weck, P. K., and Apperson, S.,** Structural and biological properties of purified bacteria-derived human fibroblast interferon, *Proc. Natl. Acad. Sci. U.S.A.*, 1983, in press.

52. **Shepard, H. M., Leung, D., Stebbing, N., and Goeddel, D. V.,** A single amino acid change in IFN-β_1 abolishes it antiviral activity, *Nature (London)*, 294, 563, 1981.

53. **Masucci, M. G., Szigeti, R., Klein, E., Gruest, J., Taira, H., Hall, A., Nagata, S., and Weissmann, C.,** Effect of interferon-alpha 1 from *E. coli* on some cell functions, *Science*, 209, 1431, 1980.

54. **Lee, S. H., Kelley, S., Chiu, H., and Stebbing, N.,** Stimulation of natural killer cell activity and inhibition of proliferation of various leukemic cells by purified human leukocyte interferon subtypes, *Cancer Res.*, 42, 1312, 1982.

55. **Herberman, R. B., Ortaldo, J. R., Mantovani, A., Hobbs, D. S., Kung, H. S., and Pestka, S.,** Effect of human recombinant interferon on cytotoxic activity of natural killer (NK) cells and monocytes, *Cell. Immunol.*, 67, 160, 1982.

56. **Targan, S. and Stebbing, N.,** *In vitro* interactions of purified cloned human interferons on NK cells: enhanced activation, *J. Immunol.*, 129, 934, 1982.

57. **Shalaby, M. R. and Weck, P. K.,** Bacteria-derived human leukocyte interferon subtypes alter *in vitro* humoral and cellular immune responses, *Cell Immunol.*, 82, 269, 1983.

58. **Torrence, P. F.,** Molecular foundations of interferon action, *Molec. Aspects Med.*, 5, 129, 1982.

59. **Weck, P. K., Apperson, S., May, L., and Stebbing, N.,** Comparison of the antiviral activity of various cloned human interferon-α subtypes in mammalian cell cultures, *J. Gen. Virol.*, 57, 233, 1981.

60. **Yelverton, E., Leung, D., Weck, P. K., Gray, P. W., and Goeddel, D. V.,** Bacterial synthesis of a novel human leukocyte interferon, *Nucleic. Acids Res.*, 9, 731, 1981.

61. **Lee, S. H., Weck, P. K., Moore, J., Chen, S., and Stebbing, N.,** Pharmacological comparison of two hybrid recombinant DNA derived human leukocyte interferons, in *Chemistry and Biology of Interferons, UCLA Symp. Mol. Cell Biol.*, Vol. 25, Merigan, T., Friedman, R., and Fox, C. F., Eds., Academic Press, New York, 1981, 341.

62. **Fish, E. N., Banerjee, K., and Stebbing, N.,** Human leukocyte interferon subtypes have different antiproliferative activities on human cells, *Biochem. Biophys. Res. Commun.*, 112, 537, 1983.

63. **Samuel, C. E. and Knutson, G. S.,** Mechanism of interferon action: cloned human leukocyte interferons induce protein kinase and inhibit Vesicular Stomatitis virus but not Reovirus replication in human amnion cells, *Virol.*, 114, 301, 1981.

64. **Weck, P. K., Apperson, S., Stebbing, N., Gray, P. W., Leung, D., Shepard, H. M., and Goeddel, D. V.,** Antiviral activities of hybrids of two major leukocyte interferons, *Nucleic Acids Res.*, 9, 6153, 1981.

65. **Weck, P. K., Rinderknecht, E., Estell, D. A., and Stebbing, N.,** Antiviral activity of bacteria derived human leukocyte interferons against encephalomyocarditis virus infection of mice, *Infect. Immun.*, 35, 660, 1982.

66. **Stebbing, N., Weck, P. K., Fenno, J. T., Apperson, S., and Lee, S. H.,** Comparison of the biological properties of natural and recombinant DNA derived human interferons, in *The Biology of the Interferon System*, DeMaeyer, E., Galasso, G., and Schellekens, H., Eds., Elsevier/North Holland, Amsterdam, 1981, 25.

67. **Braude, I. A., Lin, L. S., and Stewart, W. E.,** Isolation of a biologically active fragment of human alpha interferon, *J. Interferon Res.*, 1, 245, 1981.

68. **Borden, E. C. and Ball, L. A.,** Interferons: biochemical, cell growth inhibitory, and immunological effects, *Prog. Hematol.*, 12, 299, 1982.

69. **Salmon, S. E., Durie, B. G. M., Young, L., Lin, R. M., Trown, P. W., and Stebbing, N.,** Effects of cloned human leukocyte interferons in the human tumor stem cell assay, *J. Clin. Oncol.*, 1, 217, 1983.

70. **Epstein, L., Shen, J-T., Abele, J. S., and Reese, C. C.,** Sensitivity of human ovarian carcinoma cells to interferon and other antitumor agents as assessed by an *in vitro* semi-solid agar technique, *Ann. N.Y. Acad. Sci.*, 350, 228, 1980.

71. **Taylor-Papadimitriou, J., Shearer, M., Balkwill, F. R., and Fantes, K. H.,** Effects of the IFN-α2 and the IFN-α (Namalva) on breast cancer cells grown in culture and as xenografts in the nude mouse, *J. Interferon Res.*, 2, 479, 1982.

72. **Evinger, M., Maeda, S., and Pestka, S.,** Recombinant human leukocyte interferon produced in bacteria has antiproliferative activity, *J. Biol. Chem.*, 256, 2113, 1982.

73. **Blalock, J. E., Georgiades, J. A., Langford, M. P., and Johnson, H. M.,** Purified human immune interferon has more potent anti-cellular activity than fibroblast or leukocyte interferon, *Cell. Immunol.*, 49, 390, 1980.

74. **Freidman, R. M., Lee, G., Shifrin, S., Ambesi-Impiombato, S., Epstein, D., Jacobsen, H., and Kohn, L. D.,** Interferon interactions with thyroid cells, *J. Interferon Res.*, 2, 387, 1982.

75. **Aguet, M., Gresser, I., Hovanessian, A. G., Bandu, M.-T., Blanchard, B., and Blangy, D.,** Specific high-affinity binding of ^{125}I-labelled mouse interferon to interferon resistant embryonal carcinoma cells *in vitro, Virology,* 114, 585, 1981.

76. **Aguet, M.,** High-affinity binding of ^{125}I-labelled mouse interferon to a specific cell surface receptor, *Nature (London),* 284, 459, 1980.

77. **Branca, A. A. and Baglioni, C.,** Evidence that types I & II interferons have different receptors, *Nature (London),* 294, 768, 1981.

78. **Zoon, K., Nedden, D. Z., and Arnhules, H.,** Specific binding of human α interferon to a high affinity cell surface binding site on bovine kidney cells, *J. Biol. Chem.,* 257, 4695, 1982.

79. **Czarniecki, C. W., Fennie, C. W., Powers, D. B., and Esteu, D. A.,** Synergistic Antiviral and anti-proliferative activities of *Esherichia coli* — derived human alpha, beta, and gamma interferons., *J. Virol.,* 49, 490, 1984.

80. **Paucker, K., Dalton, B. J., Ogburn, C. A., and Torma, E.,** Multiple active sites on human interferons, *Proc. Natl. Acad. Sci. U.S.A.,* 72, 4587, 1975.

81. **Chany, C.,** Membrane-bound interferon specific cell receptor systems: role in the establishment and amplification of the antiviral state, *Biomedicine* (Paris), 24, 148, 1976.

82. **Streuli, M., Hall, A., Boll, W., Stewart, W. E., Nagata, S., and Weissmann, C.,** Target cell specificity of two species of human interferon-α produced in *Escherichia coli* and of hybrid molecules derived from them, *Proc. Natl. Acad. Sci. U.S.A.,* 78. 2848, 1981.

83. **Ankel, H., Krishnamurti, C., Besancon, F., Stefanos, S., and Falcoff, E.,** Mouse fibroblast (type I) and immune (type II) interferons: pronounced differences in affinity for gangliosides and in antiviral and antigrowth effects on mouse leukemia L1210R cells, *Proc. Natl. Acad. Sci. U.S.A.,* 77, 2528, 1980.

84. **Chandrabose, K. A., Cuatrecasas, P., Pottathil, R., and Lang, D. J.,** Interferon-resistant cell line lacks fatty acid cyclooxygenase activity, *Science,* 212, 329, 1981.

85. **Pottathil, R., Chandrabose, K. A., Cuatrecasas, P., and Lang, D. J.,** Establishment of the interferon-mediated antiviral state: role of fatty acid cyclooxygenase, *Proc. Natl. Acad. Sci. U.S.A.,* 77, 5437, 1980.

86. **Chandrabose, K. A., Cuatrecasas, P., and Pottathil, R.,** Changes in fatty acyl chains of phospholipids induced by interferon in mouse sarcoma S-180 cells, *Biochem. Biophys. Res. Commun.,* 98, 661, 1981.

87. **Chany, C.,** Reorganization of the cytoskeleton by interferon in MSV-transformed cells, *J. Interferon Res.,* 1, 323, 1981.

88. **Pfeffer, L. M. and Tamm, I.,** Effects of β interferon on concanavalin A binding and size of HeLa cells, *J. Interferon Res.,* 2, 431, 1982.

89. **Pfeffer, L. M., Wang, E., and Tamm. I.,** Interferon inhibits the redistribution of cell surface components, *J. Exp. Med.,* 152, 469, 1980.

90. **Brouty-Boye, D. and Tovey, M. C.,** Inhibition by interferon of thymidine uptake in chemostat cultures of L1210 cells, *Intervirology,* 13, 243, 1977.

91. **Gewert, D. R., Shah, S., and Clemens, M. J.,** Inhibition of cell division by interferons: changes in transport and intracellular metabolism of thymidine in human lymphoblastoid (Daudi) cells, *Eur. J. Biochem.,* 116, 487, 1981.

92. **Lindahl, P., Gresser, I., Leary, P., and Tovey, M.,** Interferon treatment of mice: enhanced expression of histocompatibility antigens on lymphoid cells, *Proc. Natl. Acad. Sci. U.S.A.,* 73, 1284, 1976.

93. **Heron, I., Hokland, M., and Berg, K.,** Enhanced expression of β_2-microglobulin and HLA antigens on human lymphoid cells by interferon, *Proc. Natl. Acad. Sci. U.S.A.,* 75, 6215, 1978.

94. **Wallach, D., Fellous, M., and Revel, M.,** Preferential effect of γ interferon on the synthesis of HLA antigens and their mRNAs in human cells, *Nature (London),* 299, 833, 1982.

95. **Knight, E. and Korant, B. D.,** A cell surface alteration in mouse L cells induced by interferon, *Biochem. Biophys. Res. Commun.,* 74, 707, 1977.

96. **Grollman, E. F., Lee, G., Ramos, S., Lazo, P. S., Koback, H. R., Friedman, R. M., and Kohn, L. D.,** Relationships of the structure and function of the interferon receptors and establishment of the antiviral state, *Cancer Res.,* 38, 4172, 1978.

97. **Kushnaryov, V. M., Sedmak, J. J., Bendler, J. W., and Grossberg, S. E.,** Ultrastructural localization of interferon receptors on the surfaces of cultured cells and erythrocytes, *Infect. Immun.,* 36, 811, 1982.

98. **Blalock, J. E. and Stanton, J. D.,** Common pathways of interferon and hormonal action, *Nature (London),* 283, 406, 1980.

99. **Blalock, J. E. and Harp, C.,** Interferon and adrenocorticotropic hormone induction of steroidogenesis, mitogenesis and antiviral activity, *Arch. Virol.,* 67. 45, 1981.

100. **Blalock, J. E. and Smith, E. M.,** Human leukocyte interferon: structural and biological relatedness to adrenocorticotropic hormone and endorphins, *Proc. Natl. Acad. Sci. U.S.A.,* 77, 5972, 1980.

101. **Blalock, J. E. and Smith, E. M.,** Human leukocyte interferon (HuIFN-α): potent endorphin-like opioid activity, *Biochem. Biophys. Res. Commun.,* 101, 472, 1981.

102. **Smith, E. M. and Blalock, J. E.,** Human lymphocyte production of corticotropin and endorphin-like substance: association with leukocyte interferon, *Proc. Natl. Acad. Sci. U.S.A.,* 78, 7530, 1981.

103. **Epstein, L. B., Rose, M. E., McManus, N. H., and Hao, C. H.,** Absence of functional and structural homology of natural and recombinant human leukocyte interferon (IFN-alpha) with human alpha-ACTH and beta-endorphin, *Biochem. Biophys. Res. Commun.,* 104, 341, 1982.

104. **Wetzel, R., Levine, H. L., Hagman, J., and Ramachandran, J.,** Human leukocyte interferon has no structural or biological relationship to corticotropin, *Biochem. Biophys. Res. Commun.,* 104, 944, 1982.

105. **Epstein, L. B.,** The effects of interferons on the immune response *in vitro* and *in vivo,* in *Interferons and Their Actions,* Stewart, W. E., II, Ed., CRC Press, Boca Raton, Fla., 1977, 91.

106. **Vilcek, J., Gresser, I., and Merigan, T. C.,** Regulatory functions of interferons, *Ann. N.Y. Acad. Sci.,* 350, 1, 1980.

107. **DeMaeyer, E. and DeMaeyer-Guiguard, J.,** Interferons as regulatory agents of the immune system, *CRC Crit. Rev. Immunol.,* 2, 167, 1981.

108. **Catalona, W. J., Ratliff, M. L., and McCool, R. E.,** Gamma interferon induced by Staphylococcus aureus protein A augments natural killing and antibody-dependent cell-mediated cytotoxicity, *Nature (London),* 291, 77, 1981.

109. **Zarling, J. M.,** Enhancement of human cytotoxic T cell responses and NK cell activity by purified fibroblast interferon and polynucleotide inducers of interferon, *Prog. Cancer Res. Ther.,* 19, 167, 1981.

110. **Herberman, R. B., Ortaldo, J. R., Rubinstein, M., and Pestka, S.,** Augmentation of natural and antibody-dependent cell-mediated cytotoxicity by pure human leukocyte interferon, *J. Clin. Immunol.,* 1, 149, 1981.

111. **Cooper, H. L., Fagnani, R., Condon, J., Trepel, J., and Lester, E. P.,** Effect of interferons on protein synthesis in human lymphocytes: enhanced synthesis of eight specific peptides in T cells and activation-dependent inhibition of overall protein synthesis, *J. Immunol.,* 128, 828, 1982.

112. **Cooper, H. L.,** Effect of bacterially produced interferon-alpha 2 on synthesis of specific peptides in human peripheral lymphocytes, *FEBS Lett.,* 140, 109, 1982.

113. **Lotzov'a, E., Savary, C. A., Gutterman, J. U., and Hersh, E. M.,** Modulation of natural killer cell-mediated cytotoxicity by partially purified and cloned interferon-alpha, *Cancer Res.,* 42, 2480, 1982.

114. **Maluish, A. E., Ortaldo, J. R., Sherwin, S., Leavitt, R., Strong, D. M., Fine, S., Wiernik, P., Oldham, R. K., and Herberman, R. B.,** Immunological monitoring of patients receiving recombinant leukocyte A interferon, *J. Cell Biochem.* Suppl. 6, 104, 1982.

115. **Gutterman, J. U., Fine, S., Quesada, J., Horning, S. J., Levine, J. F., Alexanian, R., Bernhardt, L., Kramer, M., Spiegel, H., Colburn, W., Trown, P., Merigan, T., and Dziewanowska, Z.,** Recombinant leukocyte A interferon: pharmacokinetics single-dose tolerance, and biologic effects in cancer patients, *Ann. Intern. Med.,* 96, 549, 1982.

116. **Heideman, E., Reithmann, U., Wilms, K., Treuner, J., and Niethammer, D.,** Effect of human fibroblast interferon on natural killer cell activity: stimulation *in vitro* and inhibition *in vivo, Klin. Wochenschr.,* 60, 625, 1982.

117. **Moore, M., White, W. J., and Potter, M. R.,** Modulation of target cell susceptibility to human natural killer cells by interferon, *Int. J. Cancer,* 25, 565, 1980.

118. **Welsh, R. M., Karre, K., Hansson, M., Kunkel, L. A., and Kiessling, R. W.,** Interferon mediated protection of normal and tumor target cells against lysis by mouse natural killer cells, *J. Immunol.,* 126, 219, 1981.

119. **Hokland, P. and Ellegaard, J.,** Immunological studies in chronic lymphocytic leukemia. II. Natural killer and antibody-dependent cellular cytotoxicity potentials of malignant and non-malignant lymphocyte subsets and the effect of α-interferon, *Leukemia Res.,* 5, 349, 1981.

120. **Wallach, D.,** Regulation of susceptibility of natural killer cells' cytotoxicity and regulation of HLA synthesis: differing efficacies of alpha, beta, and gamma interferons, *J. Interferon Res.,* 2, 319, 1982.

121. **Herberman, R. B. and Ortaldo, J. R.,** Natural killer cells: their role in defenses against disease, *Science,* 214, 24, 1981.

122. **Warner, J. F. and Dennert, G.,** Effects of a cloned cell line with NK activity on bone marrow transplants, tumor development and metastasis *in vivo, Nature (London),* 300, 31, 1982.

123. **Choi, Y. S., Liu, K. H., and Saunders, F. S.,** Effect of interferon alpha on pokeweed mitogen-induced differentiation of human peripheral blood B lymphocytes, *Cell. Immunol.,* 64, 20, 1982.

124. **Heron, I., Berg, K., and Cantell, K.,** Regulatory effect of interferon on T cells *in vitro, J. Immunol.,* 117, 1370, 1976.

125. **Peri, G., Palentarutti, N., Sessa, C., Mangioni, C., and Mantovani, A.,** Tumoricidal activity of macrophages isolated from human ascitic and solid ovarian carcinomas: augmentation by interferon, lymphokines, and endotoxin, *Int. J. Cancer,* 28, 143, 1981.

126. **Harfast, B., Huddlestone, J. R., Casali, P., Merigan, T. C., and Oldstone, M. B. A.,** Interferon acts directly on human B lymphocytes to modulate immunoglobulin synthesis, *J. Immunol.,* 127, 2146, 1981.

127. **Fleisher, T. A., Atallah, A. M., Tosato, G., Blaese, R. M., and Greene, W. C.,** Interferon-mediated inhibition of human polyclonal immunoglobulin synthesis, *J. Immunol.,* 129, 1099, 1982.

128. **Kadish, A. S., Tausey, F. A., Yu, G. S. M., Doyle, A. T., and Bloom, B. R.,** Interferon as a mediator of human lymphocyte suppression, *J. Exp. Med.,* 151, 637, 1980.

129. **Atallah, A. M. and Folks, T.,** Interferon enhanced human natural killer and antibody dependent cell-mediated cytotoxicity, *Int. Arch. Allergy Appl. Immunol.,* 60, 191, 1979.

130. **Parker, M. A., Mandel, A. D., Wallace, J. H., and Sonnenfeld, G.,** Modulation of human *in vitro* antibody response by human leukocyte interferon preparations, *Cell. Immunol.,* 58, 464, 1981.

131. **Jett, J. R., Mantovani, A., and Herberman, R. B.,** Augmentation of human monocyte-mediated cytolysis by interferon, *Cell. Immunol.,* 54, 425, 1980.

132. **Schultz, R. M.,** Synergistic activation of macrophages by lymphokine and lipopolysaccharide: evidence for lymphokine as the primer and interferon as the trigger, *J. Interferon Res.,* 2, 459, 1982.

133. **Schultz, R. M.,** Macrophage activation by interferons, *Lymphokine Rep.,* 1, 63, 1980.

134. **Jones, C. M., Varesio, L., Herberman, R. B., and Pestka, S.,** Interferon activates macrophages to produce plasminogen activator, *J. Interferon Res.,* 2, 377, 1982.

135. **Auget, M., Vignaux, F., Friedman, W. H., and Gresser, I.,** Enhancement of Fc gamma receptor expression in interferon-treated mice, *Eur. J. Immunol.,* 11, 926, 1981.

136. **Knop, J., Stremmer, R., Neuman, C., DeMaeyer, E., and Macher, E.,** Interferon inhibits the suppressor T cell response of delayed-type hypersensitivity, *Nature (London),* 297, 757, 1982.

137. **Fradelizi, D. and Gresser, I.,** Interferon inhibits the generation of allospecific suppressor T lymphocytes, *J. Exp. Med.,* 155, 1610, 1982.

138. **Guyre, P. M., Marganelli, P. M., and Miller, R.,** Recombinant immune interferon increases immunoglobin G FC receptors on cultured human mononuclear phagocytes., *J. Clin. Invest.,* 72, 393, 1983.

139. **Guyre, P. M., Bodwell, J. E., and Munck, A.,** Augmentation of human monocyte Fc receptors, *Nature (London),* (submitted), 1982.

140. **Renton, K. W. and Mannering, G. J.,** Repression of the hepatic cytochrome P-450 mono-oxygenase system by administered tilorone (2,7-bis[2-(diethylamino) ethoxy] fluroan-9-one dihydrochloride), *Drug Metab. Dispos.,* 4, 223, 1976.

141. **Chang, K. C., Laver, B. A., Bell, T. D., and Chai, H.,** Altered theophylline pharmacokinetics during acute respiratory viral illness, *Lancet,* 1, 1132, 1978.

142. **Mannering, G. J., Renton, K. W., ElAzhary, R., and Deloria, L.,** Effects of interferon-inducing agents on hepatic cytochrome P-450 drug metabolizing systems, *Ann. N.Y. Acad. Sci.,* 350, 314, 1980.

143. **Renton, K. W.,** Depression of hepatic cytochrome P-450-dependent mixed function oxidases during infection with encephalomyocarditis virus, *Biochem. Pharmacol.,* 30, 2333, 1981.

144. **Singh, G., Renton, K. W., and Stebbing, N.,** Homogeneous interferon from *E. coli* depresses hepatic cytochrome P-450 and drug biotransformation, *Biochem. Biophys. Res. Commun.,* 106, 1256, 1982.

145. **Moore, J. A., Marafino, B. J., and Stebbing, N.,** Influence of various purified interferons on effects of drugs in mice, *Res. Commun. Chem. Path. Pharm.,* 39, 113, 1983.

146. **Sonnenfeld, G. Harned, C. L., Thaniyavarn, S., Huoff, T., Mandel, A. D., and Nerland, D. E.,** Type II interferon induction and passive transfer depress murine cytochrome P-450 drug metabolism system, *Antimicrob. Agents Chemother.,* 17, 969, 1980.

147. **Harned, C. L., Nerland, D. E., and Sonnenfeld, G.,** Effects of passive transfer and induction of gamma (type II, Immune) interferon preparations on the metabolism of diphenyl hydrantoin by murine cytochrome P-450, *J. Interferon Res.,* 2, 5, 1982.

148. **Goeddel, D. V., Yelverton, E., Ullrich, A., Heyneker, H. L., Miozzari, G., Holmes, H., Seeburg, P. H., Dull, T., May, L., Stebbing, N., Crea, R., Maeda, S., McCandliss, R., Sloma, A., Tabor, J. M., Gross, M., Familletti, P. C., and Pestka, S.,** Human leukocyte interferon produced by *E. coli* is biologically active, *Nature* (London), 287, 411, 1980.

149. **Schellekens, H., de Reus, A., Bolhuis, R., Fountoulakis, M., Schein, C., Ecsodi, J., Nagata, S., and Weissmann, C.,** Comparative antiviral efficiency of leukocyte and bacterially produced human α-interferon in rhesus monkeys, *Nature* (London), 292, 775, 1981.

150. **Stebbing, N., Weck, P. K., Fenno, J. T., Estell, D. A., and Rinderknecht, E.,** Antiviral effects of bacteria derived human leukocyte interferons against encephalomyocarditis virus infection of squirrel monkeys, *Arch. Virol.,* 76, 365, 1983.

151. **Weck, P. K., Harkins, R. N., and Stebbing, N.,** Antiviral effects of human fibroblast interferon from *E. coli* against encephalomyocarditis virus infection of squirrel monkeys, *J. Gen. Virol.,* 64, 415, 1983.

152. **Billiau, A., DeSomer, P., Edy, V. G., DeClercq, E., and Heremens, H.,** Human fibroblast interferon for clinical trials: pharmacokinetics and tolerability in experimental animals and humans, *Antimicrob. Agents Chemother.,* 16, 56, 1979.

153. **Schellekens, H., van Eerd, P. M. C. A., de Reus, A., Weck, P. K., and Stebbing, N.,** Antiviral and side effects of interferons produced by recombinant DNA techniques as tested in rhesus monkeys, *Antiviral Res.,* 2, 313, 1982.

154. **Weimar, W., Stitz, L., Billiau, A., Cantell, K., and Schellekens, H.,** Prevention of vaccinia lesions in rhesus monkeys by human leukocyte and fibroblast interferon, *J. Gen. Virol.,* 48, 25, 1980.

155. **Knight, E. and Fahey, D.,** Human interferon beta: effects of deglycosylation, *J. Interferon Res.,* 2, 421, 1982.

156. **Stebbing, N., Dawson, K. M., and Lindley, I. J. D.,** Requirement for macrophages for interferon to be effective against encephalomyocarditis virus infection in mice, *Infect. Immunol.,* 19, 5, 1978.

157. **Schellekens, H., Weimar, W., Cantell, K., and Stitz, L.,** Antiviral effect of interferon *in vivo* may be mediated by the host, *Nature* (London), 278, 742, 1979.

158. **Smolin, G., Stebbing, N., Friedlaender, M., Friedlaender, R., and Okumoto, M.,** Natural and cloned human leukocyte interferon in herpes virus infections of rabbit eyes, *Arch. Ophthalmol.,* 100, 481, 1982.

159. **Grabner, G., Smolin, G., Okumoto, M., and Stebbing, N.,** Antiviral effects of highly purified bacteria-derived human leukocyte interferons (subtypes A and D) and a human fibroblast interferon against herpes virus infection of the rabbit eye, *Curr. Eye Res.,* 2, 785, 1983.

160. **Thacore, H. R., Mount, D. T., and Chadha, K. C.,** Interferon system of human cornea cells: interferon production, characterization, and development of antiviral state, *J. Interferon Res.,* 2, 401, 1982.

161. **Gresser, I., Maury, C., and Brouty-Boye, D.,** Mechanism of the antitumor effect of interferon in mice, *Nature* (London), 239, 167, 1972.

162. **Lee, S. H., Chiu, H., Rinderknecht, E., Sabo, W., and Stebbing, N.,** Importance of treatment regimen of interferon as an antitumor agent, *Cancer Res.,* 43, 4172, 1983.

163. **Daniel, M. D., Tamulevich, R., Bekesi, J. G., King, N. W., Falk, L. A., Silva, D., and Holland, J. F.,** Selective antiviral activity of human interferons on primate oncogenic and neurotropic herpes viruses., *Int. J. Cancer,* 27, 113, 1982.

164. **Laufs, R., Steinke, H., Jacobs, C., Hilfenhaus, J., and Karges, H.,** Influence of interferon on the replication of oncogenic herpesvirus in tissue cultures and in nonhuman primates, *Med. Microbiol. Immunol.,* 160, 285, 1974.

165. **Rabin, H., Adamson, R. H., Neubauer, R. H., Cicmanec, J. L., and Wallen, W. C.,** Pilot studies with human interferon in *herpesvirus saimiri* induced lymphoma in owl monkeys, *Cancer Res.,* 36, 715, 1976.

166. **Reid, L. M., Minato, N., Gresser, I., Holland, J., Kadish, A. R., and Bloom, B. R.,** Influence of anti-mouse interferon serum on the growth and metastasis of virus persistently-infected tumor cells and human prostatic tumors in nude mice, *Proc. Natl. Acad. Sci. U.S.A.,* 78, 1171, 1981.

167. **Yokota, Y., Kishida, T., Esaki, K., and Kawamata, J.,** Antitumor effects of interferon on transplanted tumors in congenically athmyic nude mice, *Biken J.,* 19, 125, 1976.

168. **Horoszewicz, J. S., Leong, S. S., Ito, M., Buffett, R. F., Karakovsis, C., Holyoke, E., Job, L., Dolen, J. G., and Carter, W. A.,** Human fibroblast interferon in human neoplasia: clinical and laboratory study, *Cancer Treat. Rep.,* 62, 1899, 1978.

169. **Balkwill, F., Taylor-Papadimitriou, J., Fantes, K. H., and Sebesteny, A.,** Human lymphoblastoid interferon can inhibit the growth of human breast cancer xenografts in athymic (nude) mice, *Eur. J. Cancer,* 16, 569, 1980.

170. **Kaplan, N. O. and Slimmer, S.,** Anticellular activity of human interferons, in *Cellular Responses to Molecular Modulators,* Mozes, L. and Schultz, J., Eds., Academic Press, New York, 1981, 413.

171. **Vilcek, J., Sulea, I. T., Zerebeckyj, I. L., and Yip, Y. K.,** Pharmacokinetic properties of human fibroblast and leukocyte interferon in rabbits, *J. Clin. Microbiol.,* 11, 102, 1980.

172. **Hanley, D. F., Wiranowska-Stewart, M., and Stewart, W. E.,** Pharmacology of interferons. I. Pharmacologic distinctions between human leukocyte and fibroblast interferons, *Int. J. Immunopharmacol.,* 1, 219, 1979.

173. **Bocci, V.,** Pharmacokinetic studies on interferons, *Pharmacol. Ther.,* 13, 421, 1981.

174. **Strander, H., Cantell, K., Carlstrom, G., and Jakobsson, P. A.,** Systemic administration of potent interferon to man, *J. Natl. Cancer Inst.,* 51, 733, 1973.

175. **Cantell, K. and Pyahala, L.,** Circulating interferon in rabbits after administration of human interferon by different routes, *J. Gen. Virol.,* 20, 97, 1973.

176. **Cantell, K., Pyahala, L., and Strander, H.,** Circulating human interferon after intramuscular injection into animals and man, *J. Gen. Virol.,* 22, 453, 1974.

177. **Horning, S. J., Levine, I. D., Miller, R. A., Rosenberg, S. A., and Merigan, T. C.,** Clinical and immunobiological effects of recombinant leukocyte A interferon in eight patients with advanced cancer, *JAMA,* 247, 1718, 1982.

178. **Edy, V. G., Billiau, A., and DeSomer, P.,** Comparison of rates of clearance of human fibroblast and leukocyte interferon from the circulatory system of rabbits, *J. Infect. Dis.,* 133, A18, 1976.

179. **Horoszewicz, J. S., Leong, S. S., Dolen, J. G., Holyoke, E. D., Karatousis, C., Nemoto, T., Wajsman, L. Z., Freeman, A. I., Aungst, C. W., Henderson, E., and Carter, W. A.,** Purified human fibroblast interferon and neoplasia: pharmacokinetic studies and selective properties *in vivo, J. Clin. Hematol. Oncol.,* 9, 296, 1979.

180. **Hilfenhaus, J., Damm, H., Hofstaetter, T., Mauler, R., Ronneberger, H., and Weinmann, E.,** Pharmacokinetics of human interferon-beta in monkeys, *J. Interferon Res.,* 1, 427, 1981.

181. **Billiau, A., Heremens, H., Ververken, D., VanDamme, J., Carton, H., and DeSomer, P.,** Tissue distribution of human interferons after exogenous administration in rabbits, monkeys, and mice, *Arch. Virol.,* 68, 19, 1981.

182. **Mogensen, K. E., Pyahala, L., Torma, E., and Cantell, K.,** No evidence for carbohydrate moiety affecting clearance of circulating human leukocyte interferon in rabbits, *Acta Pathol. Microbiol. Scand. Sect. B,* 82, 305, 1974.

183. **Bose, S. and Hickman, J.,** Role of the carbohydrate moiety in determining the survival of interferon in the circulation, *J. Biol. Chem.,* 252, 8336, 1977.

184. **Rohatiner, A. Z. S., Balkwill, F., Malpas, J. S., and Lister, T. A.,** A phase I study of human lymphoblastoid interferon, *Proc. Am. Soc. Clin. Onc.,* 17, 371, 1982.

185. **Tokazewski-Chen, S. A., Marafino, B. J., and Stebbing, N.,** The effects of nephrectomization on the pharmacokinetics of various cloned human interferons in the rat, *J. Pharm. Expt. Ther.,* 227, 9, 1983.

186. **Heremans, H., Billiau, A., and DeSomer, P.,** Interferon in experimental viral infections in mice: tissue interferon levels resulting from the virus infection and from exogenous interferon therapy, *Infect. Immun.,* 30, 513, 1980.

187. **Habif, D. V., Lipton, R., and Cantell, K.,** Interferon crosses blood-cerebrospinal fluid barrier in monkeys, *Proc. Soc. Exp. Biol. Med.,* 149, 287, 1975.

188. **Stringfellow, D. A.,** Murine leukemia depresses response to interferon induction correlated with a serum hyporeactive factor, *Infect. Immun.,* 13, 392, 1976.

189. **Stringfellow, D. A.,** Hyporeactivity to interferon induction: characterization of a hyporeactive factor in the serum of encephalomyocarditis virus-infected mice, *Infect. Immun.,* 11, 294, 1975.

190. **Cesario, T., Vazin, N., Slater, L., and Tiles, J.,** Inactivators of fibroblast interferon found in human serum, *Infect. Immun.,* 24, 851, 1979.

191. **Fleischmann, W. R., Lefkowitz, E. J., Georgiades, J. A., and Johnson, H. M.,** Production and function of an inhibitor of interferon action found in mouse lymphokine preparations, in *Interferon: properties and clinical uses,* Khan, A., Hill, N. O., and Dorn, G. C., Eds., Leland Fikes Foundation Press, Dallas, 1980, 195.

192. **Iwasaki, T. and Nozima, T.,** Defense mechanisms against primary influenza virus infection in mice. I. The roles of interferon and neutralizing antibodies and thymus dependence of interferon and antibody production, *J. Immunol.,* 118, 256, 1977.

193. **Svedersky, L. P., Hui, A., May, L., McKay, P., and Stebbing, N.,** Induction and augmentation of mitogen-induced immune interferon production in human peripheral blood lymphocytes by N^α-desacetyl-thymosin α_1, *Eur. J. Immunol.,* 12, 244, 1982.

194. **Huang, K. Y., Kind, P. D., Jagoda, E. M., and Goldstein, A. L.,** Thymosin treatment modulates production of interferon, *J. Interferon Res.,* 1, 411, 1981.

195. **Shohan, J., Eschel, I., Aboud, M., and Salzberg, S.,** Thymic hormonal activity on human peripheral blood lymphocytes *in vitro.* II. Enhancement of the production of immune interferon by activated cells, *J. Immunol.,* 125, 54, 1980.

196. **Lipsick, J. S. and Kaplan, N. D.,** Interleukin 2 is mitogenic for nu/nu and nu/ + murine spleen cells, *Proc. Natl. Acad. Sci. U.S.A.,* 78, 2398, 1981.

197. **Torres, B. A., Yamamoto, J. K., and Johnson, H. M.,** Interleukin 2 regulates immune interferon (IFN-γ) production by normal and suppressor cell cultures, *J. Immunol.,* 128, 2217, 1982.

198. **Johnson, H. M., Farrar, W. L., and Torres, B. A.,** Vasopression replacement of interleukin 2 requirement in gamma interferon production: lymphokine activity of a neuroendocrine hormone, *J. Immonol.,* 129, 983, 1982.

199. **Svedersky, L. P., Hui, A., Don, G., Wheeler, D., McKay, P., May, L., and Stebbing, N.,** Induction and augmentation of mitogen-induced lymphokine production in human PBL by N^α-desacetylthymosin α_1, in *Human Lymphokines,* Vol. 2, Pick, E., Ed., Academic Press, New York, 1981, 125.

200. **Henney, C. S., Kuribayashi K., Kern, D. E., and Gillis, S.,** Interleukin-2 augments natural killer cell acitivity, *Nature* (London), 291, 335, 1981.

201. **Dempsey, R. A., Dinarello, C. A., Mier, J. W., Rosenwasser, L. J., Allegretta, M., Brown, T. E., and Parkinson, D. R.,** The differential effects of human leukocytic pyrogen/lymphocyte activating factor, T-cell growth factor and interferon on human natural killer activity, *J. Immunol.,* 129, 2504, 1982.

202. **Farrar, J. J., Benjamin, W. R., Hilfiker, M. L., Howard, M., Farrar, W. L., and Fuller-Farrar, J.,** The biochemistry, biology and role of interleukin 2 in the induction of cytotoxic T cell and antibody forming B cell responses, *Immunol. Rev.,* 63, 129, 1982.

203. **Brostrom, L.-A.,** The combined effect of interferon and methotrexate on human osteosarcoma and lymphoma cell lines, *Cancer Lett.,* 10, 83, 1980.

204. **Bourgeade, M. F. and Chany, C.,** Inhibition of interferon action by cytochalasin B, colchicine, and vinblastine, *Proc. Soc. Exp. Biol. Med.*, 153, 501, 1976.

205. **Greenberg, P. L. and Mosny, S. A.,** Cytotoxic effects of interferon *in vitro* on granulolytic progenitor cells, *Cancer Res.*, 37, 1794, 1977.

206. **Chirigos, M. A. and Pearson, J. W.,** Cure of murine leukemia with drugs and interferon treatment, *Natl. Cancer Inst. Monogr.*, 51, 1367, 1973.

207. **Gresser, I., Maury, C., and Tovey, M.,** Efficacy of combined interferon cyclophosphamide therapy after diagnosis of lymphoma in AKR mice, *Eur. J. Cancer*, 14, 97, 1978.

208. **Ortaldo, J. R. and McCoy, J. L.,** Protective effects of interferon in mice previously exposed to lethal irradiation, *Radiat. Res.*, 81, 262, 1980.

209. **Levin, M. J. and Leong, P. L.,** Inhibition of human herpes viruses by combinations of acyclovir and human leukocyte interferon, *Infect. Immun.*, 32, 995, 1981.

210. **Cesario, T. C. and Slater, L. M.,** Diminished antiviral effect of human interferon in the presence of therapeutic concentrations of antineoplastic agents, *Infect. Immun.*, 27, 842, 1980.

211. **Sladek, N. E.,** Therapeutic efficacy of cyclophosphamide as a function of its metabolism, *Cancer Res.*, 32, 535, 1972.

212. **Brodeur, B. R. and Merigan, T. C.,** Suppressive effect of interferon on the humoral immune response to sheep red blood cells in mice, *J. Immunol.*, 113, 1319, 1974.

213. **Brodeur, B. R. and Merigan, T. C.,** Mechanism of the suppressive effect of interferon on antibody synthesis *in vivo*, *J. Immunol.*, 114, 1323, 1975.

214. **Johnson, H. M., Smith, B. G., and Baron, S. J.,** Inhibition of the primary *in vitro* antibody response by interferon preparations, *J. Immunol.*, 114, 493, 1975.

215. **Polak, L. and Turk, J. L.,** Reversal of immunological tolerance by cyclo-phosphamide through inhibition of suppressor cell activity, *Nature* (London), 249, 654, 1974.

216. **Fefer, A.,** Immunotherapy and chemotherapy of moloney sarcoma virus-induced tumors in mice, *Cancer Res.*, 29, 2177, 1969.

217. **Hengst, J. C. D., Mokyr, M. B., and Dray, S.,** Cooperation between cyclophosphamide tumoricidal activity and host antitumor immunity in the cure of mice bearing large MOPC-315 tumors, *Cancer Res.*, 41, 2163, 1981.

218. **Slater, L. M., Wetzel, M. W., and Cesario, T.,** Combined interferon-antimetabolite therapy of murine L1210 leukemia, *Cancer* 48, 5, 1981.

219. **O'Neill, G. J., and Stebbing, N.,** Influence of *C. parvum* on the effectiveness of passive serotherapy in the control of the EL4 lymphoma in C57B1/6 mice, *Br. J. Cancer*, 41, 243, 1980.

220. **Evans, J. R. and Johnson, H. M.,** The induction of at least two distinct types of interferon in mouse spleen cell cultures by *Corynebacterium parvum*, *Cell. Immunol.*, 64, 64, 1981.

221. **Hokland, P., Ellegaard, J., and Heron, I.,** Immunomodulation by *Corynebacterium parvum* in normal humans, *J. Immunol.*, 124, 2180, 1980.

222. **Sacks, S. L., Scullard, G. H., Pollard, R. B., Gregory, P. B., Robinson, W. S., and Merigan, T. C.,** Antiviral treatment of chronic hepatitis B virus infection. 4. Pharmacokinetics and side effects of interferon and adenine arabinoside alone and in combination, *Antimicrob. Agents. Chemother.*, 21, 93, 1982.

223. **Borgstrom, S., vonEyben, F. E., Flodgren, P., Axelson, B., and Sjogren, H. O.,** Human leukocyte interferon and cimetidine for metastatic melanoma, *N. Engl. J. Med.*, 307, 1080, 1982.

224. **Scott, G. M., Secher, D. S., Flowers, D., Bate, J., Cantell, K., and Tyrrell, D. A. J.,** Toxicity of interferon, *Br. Med. J.*, 282, 1345, 1981.

225. **Louie, A. C., Gallagher, J. C., Sikora, K., Levy, R., Rosenberg, S. A., and Merigan, T. C.,** Follow-up observations on the effect of human leukocyte interferon in non-Hodgkin's lymphoma, *Blood*, 58, 712, 1981.

226. **Gutterman, J. U., Blumenschein, G. R., Alexanian, R., Yap, H. Y., Buzdar, A. U., Cabanillas, F., Hortobagyi, G. N., Hersh, E. M., Rasmussen, S. L., Harmon, M., Kramer, M., and Petska, S.,** Leukocyte interferon-induced tumor regression in human metastatic breast cancer, multiple myeloma, and malignant lymphoma, *Ann. Intern. Med.*, 93, 399, 1980.

227. **Borden, E. C., Holland, J. F., Dao, T. L., Gutterman, J. U., Wiener, L., Chang, Y. C., and Patel, J.,** Leukocyte-derived interferon (alpha) in human breast carcinoma, *Ann. Intern. Med.*, 97, 1, 1982.

228. **Ingimarsson, S., Cantell, K., and Strander, H.,** Side effects of long-term treatment with human leukocyte interferon, *J. Infect. Dis.*, 140, 560, 1979.

229. **Scott, G. M., Wallace, J,. Tyrrell, D. A. J., Cantell, K., Secher, D. S., and Stewart, W. E.,** Interim report on studies on "toxic" effects of human leukocyte-derived interferon-alpha (HuIFN-α), *J. Interferon Res.*, 2, 127, 1982.

230. **Ezaki, K., Ogawa, M., Okabe, K., Abe, K., Inone, K., Horikoshi, N., and Inagaki, J.,** Clinical and immunological studies of human fibroblast interferon, *Cancer Chemother. Pharmacol*, 8, 47, 1982.

231. **Treuner, J., Dannecker, G., Joester, K. E., Hettinger, A., and Niethammer, D.,** Pharmacological aspects of clinical stage I/II trials with human beta interferon, *J. Interferon Res.,* 1, 373, 1981.

232. **Priestman, T. J.,** Initial evaluation of human lymphoblastoid interferon in patients with advanced malignant disease, *Lancet,* 1, 113, 1980.

233. **Greenberg, H. B., Pollard, R. B., and Lutwick, L. I.,** Effect of human leukocyte interferon on hepatitis B virus infection in patients with chronic active hepatitis, *N. Engl. J. Med.,* 275, 517, 1976.

234. **Emodi, G., Just, M., Hernandez, R., and Hirt, J. R.,** Circulating interferon in man after administration of exogenous human leukocyte interferon, *J. Nat. Cancer Inst.,* 54, 1045, 1975.

235. **Fleming, W. A., McNeill, T. A., and Killen, M.,** The effects of an inhibiting factor (interferon) on the *in vitro* growth of granulocyte-macrophage colonies, *Immunol.,* 23, 429, 1972.

236. **Verma, D. S., Spitzer, G., Gutterman, J. U., Zander, A. R., McCredie, K. B., and Dicke, K. A.,** Human leukocyte interferon preparation blocks granulopoietic differentiation, *Blood,* 54, 1423, 1979.

237. **Bocci, V.,** Possible causes of fever after interferon administration, *Biomedicine* (Paris), 32, 159, 1980.

238. **Panem, S., Check, I. J., Henricksen, D., and Vilcek, J.,** Antibodies to alpha-interferon in a patient with systemic lupus erythematosus, *J. Immunol.,* 129, 1, 1982.

239. **Vallbracht, A., Treuner, J., Manncke, K. H., and Niethammer, D.,** Autoantibodies against human beta interferon following treatment with interferon, *J. Interferon Res.,* 2, 197, 1982.

240. **Finter, N. B., Woodrooffe, J., and Priestman, T. J.,** Monkeys are insensitive to pyrogenic effects of human alpha-interferons, *Nature* (London), 298, 301, 1982.

241. **Gresser, I., Tovey, M., Maury, C., and Chouroulinkov, I.,** Lethality of interferon preparations for newborn mice, *Nature* (London), 258, 76, 1975.

242. **Gresser, I. and Bourali, C.,** Development of newborn mice during prolonged treatment with interferon, *Eur. J. Cancer,* 6, 553, 1970.

243. **Gresser, I., Maury, C., Tovey, M., Morel-Maroger, I., and Pontillon, F.,** Progressive glomerulonephritis in mice treated with interferon preparations at birth, *Nature* (London), 263, 420, 1976.

244. **Frayssinet, C., Gresser, I., Tovey, M., and Lindahl, P.,** Inhibitory effect of potent interferon preparations on the regeneration of mouse liver after partial hepatectomy, *Nature* (London), 245, 146, 1973.

245. **Sonnenfeld, G., Meruelo, D., McDevitt, H. O., and Merigen, T. C.,** Effect of type I and type II interferons on murine thymocyte surface antigen expression: induction or selection?, *Cell. Immunol.,* 57, 427, 1981.

246. **Van den Broek, A. A.,** Mechanism of immune suppression by cortisone acetate, in *Immune Suppression and Histopathology of the Immune Response,* Van den Broeck, A. A., Ed., Drukkerij van den Deren, Groningen, The Netherlands, 1971, 83.

247. **Gresser, I., Guy-Grand, D., Maury, C., and Maunoury, M. T.,** Interferon induces peripheral lymphadenopathy in mice, *J. Immunol.,* 127, 1569, 1981.

248. **Weck, P. K., Hamilton, E., Shalaby, M. R., Stebbing, N., Leung, D., Shepard, M., and Goeddel, D. V.,** Biological properties of genetic hybrids of bacteria-derived human leukocyte interferons, *J. Cell. Biochem.,* (Suppl. 6), 104, 1982.

249. **Franke, A. E., Shepard, H. M., Houck, C. M., Leung, D. W., Goeddel, D. V., and Lawn, R. M.,** Carboxy terminal region of hybrid leukocyte interferons affects antiviral specificity, *DNA,* 1, 223, 1982.

250. **Slocombe, P., Easton, A., Boseley, P., and Burke, D. C.,** High-level expression of an interferon α_2 gene cloned in phage M13mp7 and subsequent purification with a monoclonal antibody, *Proc. Natl. Acad. Sci. U.S.A.,* 79, 5455, 1982.

251. **Stebbing, N., Lee, S. H., Marafino, B. J., Weck, P. K., and Renton, K. W.,** Activity of cloned gene products in animal systems, in *From Gene to Protein, Translation into Biotechnology,* Ahmad, F., Schultz, J., and Smith, E. E., Eds., Academic Press, New York, 1982, 445.

252. **Strander, H., Cantell, K., Ingimarsson, S., Jakobsson, P. A., Nilsonne, U., and Soderberg, G.,** Interferon treatment of osteogenic sarcoma: a clinical trial, *Fogarty Int. Center Proc.,* 28, 377, 1977.

253. **Merigan, T. C., Sikora, K., Bruden, J. H., Levy, R., and Rosenberg, S. A.,** Preliminary observations on the effect of human leukocyte interferon in non-Hodgkins lymphoma, *N. Engl. J. Med.,* 299, 1449, 1978.

254. **Krown, S. E., Burik, M., Kirkwood, J. M., Kern, D., Nordlund, J. J., Morton, D. L., and Oettgen, H. F.,** Human leukocyte interferon in malignant melanoma: preliminary report of the American Cancer Society trial, *Proc. Am. Assoc. Cancer Res.,* 22, 158, 1981.

255. **Krown, S. E., Stoopler, M. B., Cunningham-Rundles, S., and Oettgen, H. F.,** Phase II trial of human leukocyte interferon in non-small cell lung cancer (NSCLC), *Proc. Am. Assoc. Cancer Res.,* 21, 179, 1980.

256. **Schouten, T. J., Weimar, W., Boss, J. H., Bos, C. E., Cremers, C. W., and Schellekens, H.,** Treatment of juvenile laryngeal papillomatosis with two types of interferon, *Laryngoscope,* 92, 686, 1982.

257. **Meyers, J. D., McGuffin, R. W., Bryson, Y. J., Cantell, K., and Thomas, E. D.,** Treatment of cytomegalovirus pneumonia after marrow transplant with combined vidarabine and human leukocyte interferon, *J. Infect. Dis.,* 146, 80, 1982.

258. **Scott, G. M., Phillpotts, R. J., Wallace, J., Secher, D. S., Cantell, K., and Tyrrell, D. A. J.,** Purified interferon as protection against rhinovirus infection, *Brit. Med. J., 284,* 1822, 1982.

259. **Scott, G. M., Wallace, J., Greiner, J., Philpotts, R. J., Gauci, C. L., and Tyrrell, D. A. J.,** Prevention of rhinovirus colds by human interferon alpha-2 from *Escherichia coli, Lancet,* 1, 186, 1982.

260. **Levin, S., Hahn, T., Rosenberg, H., and Bino, T.,** Treatment of life-threatening viral infections with interferon alpha: pharmacokinetic studies in a clinical trial, *Isr. J. Med. Sci.,* 18, 439, 1982.

261. **Weimar, W., Tenkate, F. J. P., Masurel, N., Heijtink, R. A., Schalm, S. W., and Schellekens, H.,** Double-blind study of leukocyte interferon administration in chronic HB$_S$Ag-positive hepatitis, *Lancet,* 1, 336, 1980.

262. **Merigan, T. C., Robinson, W. S., and Gregory, P. B.,** Interferon in chronic hepatitis B infection, *Lancet,* 1, 422, 1980.

263. **Pazin, G. J., Armstrong, J. A., Larn, M. T., Tam, G. C., Janetta, P. J., and Ho, M.,** Prevention of reactivated herpes simplex infection by human leukocyte interferon after operation on the trigeminal root, *N. Engl. J. Med.,* 301, 225, 1979.

264. **Merigan, T. C., Rand, K. H., Pollard, R. B., Abdullah, P. S., Jordan, G. W., and Fried, R. P.,** Human leukocyte interferon for treatment for herpes zoster in patients wich cancer, *N. Engl. J. Med.,* 298, 981, 1978.

265. **Romano, A., Revel, M., Guarari-Rotman, D., Blumentahl, M., and Stein, R.,** Use of human fibroblast-derived (beta) interferon in the treatment of epidemic adenovirus, *J. Interferon Res.,* 1, 95, 1980.

266. **Oh, J. O., Ostler, H. B., and Schaehter, J.,** Protective effect of a synthetic polynucleotide complex (poly I:C) on ocular lesions produced by trachoma agents in rabbits, *Infect. Immun.,* 1, 566, 1970.

267. **Kazar, J., Gillmore, J. D., and Gordon, I. B.,** Effect of interferon and interferon inducers on infections with a nonviral intracellular microorganism, *Chlamydia trachomatis, Infect. Immun.,* 3, 825, 1972.

268. **Ojo-amaize, E. A., Salimonu, L. S., Williams, A. I. O., Akinwolere, O. A. O., Shabo, R., Alm, G. V., and Wigzell, H.,** Positive correlation between degree of parasitemia, interferon titers, and natural killer cell activity in *Plasmodium falciparum* infected children, *J. Immunol.,* 128, 2296, 1981.

269. **Rytel, M. W., Rose, H. D., and Stewart, R. D.,** Absence of circulating interferon in patients with malaria and with American trypanosomasis, *Proc. Soc. Exp. Biol. Med.,* 144, 122, 1973.

270. **Roder, J. C., Haliotis, T., Klein, M., Koree, S., Jett, J. R., Ortaldo, J., Herberman, R. B., Katz, P., and Fauci, A. S.,** A new immunodeficiency disorder in humans involving NK cells, *Nature* (London), 284, 553, 1980.

271. **Wyler, D. J., Liang, L. F., Downey, E., and Krim, M.,** Exogenous interferon administration in experimental leishmaniasis: *in vivo* and *in vitro* studies, *Am. J. Trop. Med. Hyg.,* 31, 740, 1982.

272. **Ikic, D., Brnobic, A., Lipozencic, J., Maleck, J., Smerdel, S., Delimar, N., and Soos, E.,** Topical application of human leukocyte interferon in stasis ulcers and deep burns, *Clin. Pharmacol. Ther. Tox.,* 19, 540, 1981.

273. **Parsons, J. A.,** Endocrine pharmacology, in *Peptide Hormones,* Parsons, J. A., Ed., McGraw Hill, New York, 1979, 67.

274. **Edge, M. D., Greene, A. R., Heathcliffe, G. R., Meacock, P. A., Schuch, W., Scanlon, D. B., Atkinson, T. C., Newton, C. R., and Markham, A. F.,** Total synthesis of a human leukocyte interferon gene, *Nature* (London), 292, 756, 1981.

275. **Haller, O., Arnheiter, H., Lindenmann, J., and Gresser, I.,** Host gene influences sensitivity to interferon action selectively for influenza virus, *Nature* (London), 283, 660, 1980.

276. **Dandoy, F., DeMaeyer-Guignard, J., Bailey, D., and DeMaeyer, E.,** Mouse genes influence antiviral action of interferon *in vivo, Infect. Immun.,* 38, 89, 1982.

277. **Tan, Y. H., Schneider, E. L., Tischfield, J., Epstein, C. J., and Ruddle, F. H.,** Human chromosome 21 dosage: effect on the expression of the interferon induced antiviral state, *Science,* 186, 61, 1974.

278. **Stebbing, N.,** Protection of mice against infection with wild type Mengo virus and an interferon sensitive mutant (IS-1) by polynucleotides and interferons, *J. Gen. Virol.,* 44, 255, 1979.

279. **Zwartzky, R., Gresser, I., DeMaeyer, E., and Kirchner, H.,** The role of interferon in the resistance of C57Bl/6 mice to various doses of herpes simplex virus type I, *J. Infect. Dis.,* 146, 405, 1982.

280. **Herlyn, D. and Koprowski, H.,** IgG2a monoclonal antibodies inhibit human tumor through interaction with effector cells, *Proc. Natl. Acad. Sci. U.S.A.,* 79, 4761, 1982.

281. **Secher, D. S. and Stebbing, N.,** unpublished.

282. **Fenno, J. T., Weck, P. K., and Stebbing, N.,** unpublished results.

283. **Weck, P. K.,** unpublished.

284. **Lee, S. H., Chiu, H., and Stebbing, N.,** unpublished.

285. **Lee, S. H., Chiu, H., and Stebbing, N.,** unpublished.

286. **Stebbing, N.,** unpublished.

287. **Finter, N.,** personal communication.

288. **Fenno, J. T., Weck, P. K., and Stebbing, N.,** unpublished results.

Chapter 6

HUMAN CLINICAL TRIALS OF BACTERIA-DERIVED HUMAN α INTERFERON

Zofia E. Dziewanowska, Leon L. Bernhardt, and Seymour Fein

TABLE OF CONTENTS

I. INTRODUCTION

Interferon was described by Isaacs and Lindenman in 1957 and given its name because of its ability to interfere with viral reproduction in tissue culture.[1] In addition, it was soon noted that it could inhibit the multiplication of different viruses and some tumor cell lines both in vitro and in vivo.[2-5]

Clinial studies of interferon in cancer and viral infections were impeded by scarce supplies and the enormous expense of production. Despite these problems, a few trials were begun in the 1960s and clinical activity continued at an accelerated pace in the 1970s with buffy coat derived human leukocyte interferon produced in the laboratory of Dr. Kari Cantell in Finland. This material was available in only very limited amounts and was a crude preparation containing less than 1% interferon. Thus, most of the trials performed with Cantell interferon were limited to very small numbers of patients, utilized low doses of interferon and were selected largely on the basis of availability of material. Systematic phase I studies to determine such vital data as maximum tolerated dose, dose related immunologic effects, dose response, etc. could not be done at that time.

Nevertheless, varying degrees of antitumor activity were observed in several of these studies. Non-Hodgkins lymphoma, multiple myeloma, and breast cancer were the cancers most extensively evaluated in which some biologic activity of interferon was seen. Also various degrees of efficacy in viral diseases have been reported with parenteral administration of interferon derived from cultures of human leukocytes. Prominent among these demonstrations of usefulness of interferon were studies run in herpes zoster and in prophylaxis of cytomegalovirus (CMV) infections in renal transplant recipients.[6,7] Review of clinical experience with human cell culture derived interferons can be found elsewhere.[8] At the same time, progress in biotechnology made available recombinant DNA techniques capable of biosynthetically producing large quantities of interferon. Hybridoma-produced monoclonal antibodies with the ability to purify interferon proteins to homogeneity were developed.[9] Availability of an unlimited amount of pure leukocyte interferon obtained through genetic engineering allowed for the first time systematic evaluation of this protein in a sufficient number of patients. For the first time, the question of tolerance and effect on immune parameters, human tumors, and viral infections could be examined using the various doses and dose schedules.

The development of recombinant interferon introduced however, an additional dimension to this research. The several species of leukocyte interferons obtained posed the challenge to scientists to determine whether these subspecies would exhibit a therapeutic specificity for a particular tumor type or virus.[8]

In addition, because interferons exhibit various degrees of species specificity, preclinical models relevant to clinical situations are difficult to identify and interpret. Such fundamental issues as interferon mechanism of action in producing an antitumor and antiviral effect remain unresolved. If interferon primarily exerts a direct antiproliferative effect on cancer cells, the maximum tolerated dose would be expected to produce the optimal effect. This is the case with cytotoxic chemotherapeutic agents. Interferon may work, however, primarily through indirect mechanisms via the host immune systems, in which case a dose other than the maximum tolerated dose (MTD) might be optimal. This so-called maximum immunomodulatory dose-MID had to be empirically determined by careful testing of various immunologic parameters at various doses and dosage schedules. At the present time, interferon's importance and ultimate role in the treatment of cancer and the majority of viral infections remains to be defined by properly controlled studies using recombinant material in optimal dose and appropriate therapeutic regimens.

II. DESCRIPTION OF RECOMBINANT LEUKOCYTE A INTERFERON

An extensive phase I clinical program was initiated by Hoffmann-La Roche in January, 1981 with Recombinant Leukocyte A Interferon, a genetically engineered human leukocyte interferon. The 165 amino acid polypeptide with a molecular weight of 19,241 daltons is being referred to as recombinant leukocyte interferon A: rIFN-αA.[8]

Isolation, expression, and purification of rIFN-αA were done by Hoffmann-La Roche Inc. and Genentech.[8,9] The plasmid containing an entire leukocyte interferon gene was identified and engineered in *E. coli*. Purification was done using a specific monoclonal antibody column to rIFN-αA. The purified protein was homogeneous by sodium dodecyl sulfate-polyacrylamide gel electrophoresis, and was more than 95% pure. Its specific activity was approximately 2×10^8 units/mg protein when tested on AG 1732-human fibroblast or MDBK-bovine kidney-cell lines. Cultures of the material for bacteria and mycoplasma were sterile and a limulus test for endotoxin was negative.

This recombinant interferon showed equivalent in vitro antiviral and antiproliferative activity compared to crude and purified natural leukocyte interferons.[8,9] In addition, rIFN-αA stimulated natural killer cell activity in vitro, and inhibited hematopoietic colony formation.[10,11] Preclinical in vivo testing showed antiviral activity comparable to human leukocyte interferon and was safely used in rabbits and squirrel monkeys.[12] The material for clinical trials was provided in ampules of 3, 18, and 50×10^6 units. The final lyophilized preparation contained human serum albumin and was reconstituted immediately prior to use.

III. PHASE I CLINICAL STUDIES OF rIFN-αA IN CANCER PATIENTS

Because rIFN-αA was shown to possess an antiproliferative activity, the phase I evaluation of parenteral route of administration was performed mostly in cancer patients.

To date, two single-dose studies and three multiple-dose phase I studies of rIFN-αA have been completed and analyzed. All trials utilized a heterogeneous population of patients with disseminated cancers, either refractory to standard therapy or for which no effective treatment existed.

The first single-dose study was performed by Gutterman at M.D. Anderson and Merigan at Stanford. Sixteen patients received an initial intramuscular injection of 3×10^6 U of interferon. This was followed by escalating doses, with a crossover comparison between rIFN-αA and partially purified human leukocyte interferon HuIFNα prepared by Cantell.[13] The second single dose study was performed by Krown at Memorial Sloan-Kettering Cancer Center and compared the bioavailability and tolerance of intramuscular and intravenous injections of rIFN-αA at 3, 9, and 18×10^6 U doses. Three patients were studied at each dose level. The three multiple dose studies were performed at the National Cancer Institute and at Memorial Sloan-Kettering Cancer Center.[14-16] rIFN-αA was given intramuscularly at doses ranging from 0.1 to 136×10^6 U per injection on schedules of 3 times weekly, daily, and twice daily. Dose escalation occurred in sequentially entered parallel groups, usually of five patients per schedule. Duration of treatment was 28 days with provisions for continuation if an objective tumor response was observed. The objectives of these studies were to determine the maximum tolerated and optimal immunomodulatory doses of rIFN-αA on each schedule and to evaluate the pharmacokinetics and preliminary antitumor activity of this interferon. In all, 184 patients were entered, and 126 completed the 28-day course of therapy. A total of 58 patients were removed from the study: 20 due to rIFN-αA intolerance, 26 for tumor progression, and 12 due to intercurrent illness or complications of their malignancy without tumor progression.

A. Clinical Side Effects of rIFN-αA

The clinical side effects observed during the single dose studies were predictive of those found in the multiple-dose studies described below and will not be separately presented.

Table 1 depicts the most important clinical side effects of rIFN-αA as a percentage of evaluable patients experiencing each side effect. Analysis is provided for selected weekly cumulative doses from 0.3 to 408 × 10^6 U with the schedule for each dose noted.

Clinical toxicities could be categorized as acute and chronic. Acute side effects included fever and flu-like symptoms of headache, myalgias, and chills. Partial tolerance or tachyphylaxis developed to these side effects after the first few injections. A rough dose relationship in terms of frequency of side effects existed to about a dose of 18 or 36 × 10^6 U per injection above which no clear-cut dose response was noted. Chronic side effects included fatigue, anorexia, and weight loss. These seemed more related to the weekly cumulative dose and were persistent as long as rIFN-αA treatment continued. At doses of 50 × 10^6 U and greater, most patients were pretreated with acetaminophen to moderate the fever and flu-like symptoms following the first few injections. This may have confounded the analysis of an ongoing dose response relationship between higher doses of rIFN-αA and the severity of the acute clinical side effects of fever, myalgia, chills, and headache.

Schedule of administration did not dramatically affect the incidence or severity of these clinical symptoms. There was the impression of slightly better tolerance developing to the acute side effects on daily and twice daily administration and slightly less fatigue on the 3 × weekly schedule for comparable individual and weekly cumulative doses, respectively.

Fever was almost universal except at the lowest dose, beginning 2 to 4 hr after an injection, often preceded by a chill and peaking at 6 to 12 hr. It usually resolved spontaneously by 24 hr. At doses of 18 × 10^6 U and above, a maximum fever of >103°F was common. "Flu-like" symptoms tended to be associated with fever and follow a similar time course and dose relationship. Both fever and flu-like symptoms could be partially ameliorated by pretreatment with acetaminophen.

Fatigue was the most important clinical side effect noted and was potentially dose limiting. It tended to begin in the first 1 to 2 weeks of treatment and was persistent. Fatigue became more frequent as dose increased up to 63 × 10^6 U weekly at which point almost all patients noted it to some degree. It tended to become increasingly severe as doses and duration of administration increased, often causing patients to be at bedrest for much of the day at the top doses. Anorexia accompanied by the development of a 5 to 10 % or greater weight loss was seen in association with the fatigue. Other gastro-intestinal symptoms such as nausea, vomiting, and diarrhea were sporadic but did show increased incidences at doses 36 × 10^6 U and above.

Neurologic side effects included 19 patients with numbness and parasthesias of fingers and toes. In most patients this was not associated with objective findings on neurologic examination, although two patients were found to have a decrease in sensory perception of the fingertips attributed to a mild sensory neuropathy. In several patients these symptoms were quite transient lasting a few seconds to hours but in others it persisted for several days. Numbness and parasthesias often resolved despite continued administration of rIFN-α and invariably resolved with its discontinuation. Nerve conduction studies and electromyograms were done on several of these patients and were normal in all instances. Eight patients reported transient confusion or disorientation, and eight patients noted emotional lability. These manifestations were of uncertain etiology but usually occurred at higher doses.

Twelve patients developed recurrent herpes simplex labialis, usually during the first week of therapy. Seven of these occurred at doses of 50 × 10^6 U. No disseminated herpes or other viral disorders were observed. Local tolerance at the injection site for both I.M. and I.V. administration was excellent. Minimal local tenderness and induration rarely occurred.

B. Laboratory Changes in rIFN-αA Trials

As with clinical side effects, laboratory changes observed during the single dose studies were similar in type to those seen in the later studies.

Table 1
COMMONLY ENCOUNTERED CLINICAL TOXICITIES OF
rIFN-αA[a]

Weekly dose × 10⁶ U	Fever	Fatigue	GI disturbances[b]	Flu-like[c] system	Wt loss>5%
0.3 (0.1 3×w)	0	0	0	0	0
3 (1 3×w)	100	0	40	20	0
9 (3 3×w)	100	0	29	0	14
27 (9 3×w)	100	0	50	40	17
54 (18 3×w)	100	38	50	38	13
63 (9QD,4.5 BID)	100	90	60	40	40
150 (50 3×w)	100	100	100	100	50
252 (36QD,18 BID)	100	93	67	73	40
258 (86 3×w)	100	90	100	70	33
378 (54QD,27 BID)	100	92	68	72	36
408 (136 3×w)	100	100	100	100	40

[a]   Expressed as a percentage of evaluable patients experiencing toxicity.
[b]   GI disturbances include nausea, vomiting, anorexia, and diarrhea.
[c]   Flu-like symptoms include headache, myalgias, and chills.

Table 2 depicts the commonly encountered laboratory changes observed during the multiple-dose phase I studies of rIFN-αA. In general, schedule was not important. The weekly cumulative dose correlated roughly with severity. The main laboratory side effects were reflected by change in hematologic and hepatic parameters.

The frequency of neutropenia was fairly constant at weekly doses of 27 × 10⁶ U and above. Nadirs below 1000 and 500 were more common with increasing doses. The WBC decreased within 1 to 2 days following the first injection and nadired at 7 to 10 days plateauing or slightly increasing thereafter despite continued treatment with rIFN-αA. Leukopenia and neutropenia were rapidly reversible within days after discontinuation of rIFN-αA. No sepsis or other serious infections were noted during neutropenic episodes in these patients.

The platelet count was less affected by rIFN-αA. A 50% decrease from baseline was common at doses of 9 × 10⁶ U and above, but counts under 75,000 were infrequent and tended to occur mainly in patients with compromised bone marrow reserve. Recovery occurred within 1 week of discontinuation. No hemorrhagic episodes were noted. The most frequently affected parameter of liver function was transaminase elevation. The increases were only of significant degree at the very highest doses. Elevations tended to peak during the third week of treatment and returned to baseline in 10 to 14 days after discontinuation. No abnormalities of bilirubin, albumin, prothrombin time, or other hepatic synthetic function were seen.

The signs of involvement of renal functions were minimal and only occasional transient microscopic hematuria or pyuria were observed. Three patients developed mild azotemia of uncertain cause but probably related to their underlying disease, and two patients developed proteinuria of 1 g/day without casts or other signs of glomerulonephritis, which resolved with discontinuation of therapy. Because these trials were open, nonrandomized, and included patients with advanced cancer in whom some renal abnormalities might occur on the basis of underlying disease, it is difficult to be certain of the role, if any, which interferon played in these cases.

C. Pharmacokinetics of rIFN-αA

As part of the phase I evaluation of rIFN-αA, a pharmacokinetic profile was generated. Serum interferon titers were measured in antiviral units/mℓ by a bioassay using MDBK cells

Table 2
**COMMONLY ENCOUNTERED LABORATORY TOXICITIES OF
rIFN-αA[a]**

Weekly Dose × 10⁶ U	Neutrophil Count			Platelet Count		SGOT >250 U		
	<50%	<1000	<500	<50%	<75,000	2×	3×	3×
0.3 (0.1 3×w)	0	0	0	0	0	0	0	0
3 (1 3×w)	20	0	0	0	0	0	0	0
9 (3 3×w)	33	0	0	0	0	0	0	0
27 (9 3×w)	71	0	0	29	0	14	0	0
54 (18 3 3×w)	63	13	0	38	13	63	0	0
63 (9QD,4.5 BID)	90	30	0	60	20	50	30	0
150 (50 3×w)	83	17	0	17	0	17	17	0
252 (36QD,18 BID)	73	40	7	53	7	20	40	13
258 (86 3×w)	89	33	22	78	44	44	33	11
378 (54QD,27 BID)	76	40	16	68	36	20	28	36
408 (136 3×w)	75	25	0	50	0	25	50	25

[a] Expressed as a percentage of evaluable patients experiencing toxicity.

as targets and vesicular stomatitis virus. Interferon titers are expressed as reciprocals, with the dilutions producing a 50% reduction of virus cytopathic effect. All samples were corrected to the standard G023-901-527 reagent from the National Institute of Allergy and Infectious Diseases, National Institutes of Health, Bethesda, Maryland.

An enzyme-linked immunoassay using two monoclonal antibodies to rIFN-αA was also performed. The assay depends on the stoichiometric binding to interferon by two monoclonal antibodies, one of which is linked to peroxidase. The peroxidase reacts with o-phenylene-diamine in the presence of peroxide. The color produced was read in a spectrophotometer at 492 nm and the serum interferon level expressed in pg/mℓ.

Figure 1 shows the mean serum concentrations of rIFN-αA as measured by the bioassay, for the 16 patients treated in the escalating single dose study. Figure 2 shows the same data as measured by the enzyme-linked immunoassay. The pharmacokinetic profile of rIFN-αA was essentially identical in both assays.

In general, mean maximum observed serum concentration (^{C}MAX) and area under curve (AUC) values increased with increasing doses. When ^{C}MAX and AUC were normalized for the dose given and represented as pg/mℓ/10⁶ unit dose or units/mℓ/10⁶ unit dose, there was dose proportionality. The time of the maximum observed serum concentration t_{max} tended to increase slightly with increasing doses when one injection site was used and varied from 4.6 ± 2.5 hr at 3 × 10⁶ U to 6.1 ± 3.6 hr at 72 × 10⁶ units. This result is consistent with a depot effect of the intramuscular route at the injection site. The half-life ranged from 6 to 8 hr regardless of dose. rIFN-αA was not detected in the urine as measured by the enzyme immunoassay after doses ranging from 3 to 72 × 10⁶ units.

Figure 3 shows the arithmetic mean serum concentration of rIFN-αA and HuIFNα at doses of 3 × 10⁶ units and 9 × 10⁶ units respectively, as measured by bioassay. The only statistically significant difference between the rIFN-αA and HuIFNα values is the AUC at 9 × 10⁶ units, where the HuIFN-α value is significantly larger than the rIFN-αA value, $p < 0.05$. The half-lives of elimination were calculated at the 9 × 10⁶ units doses and were 7.3 and 8.2 hr for HuIFNα and rIFN-αA, respectively. Serum interferon levels were closely monitored during the multiple dose studies as well. The designs for these trials provided for only 5 patients to be entered at each dose level on each schedule and therefore did not permit full multiple dose pharmacokinetic analysis to be performed. However, interesting differences were noted among the different schedules studied.

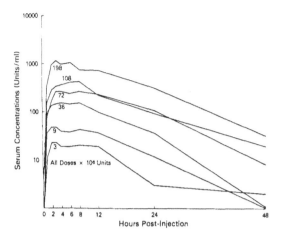

FIGURE 1. The arithmetic mean serum concentrations of interferon as measured by the bioassay with MDBK cells as target cells. The numbers of patients measured at 3, 9, 36, 72, 108, and 198 million units are 16, 16, 16, 16, 14, and 5, respectively.

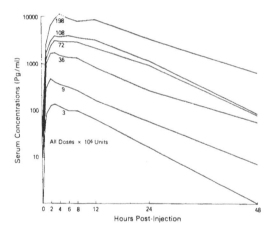

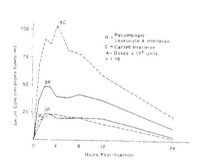

FIGURE 2. The arithmetic mean serum concentrations of interferon as measured by the enzyme immunoassay. The numbers of patients measured at 3, 9, 36, 72, 108, and 198 million units are 16, 16, 16, 16, 14, and 5, respectively.

FIGURE 3. The comparative arithmetic mean serum levels of 16 patients treated with rIFN-αA and IFN-C as measured by a modified bioassay at 3 and 9 million units.

Figure 4 depicts the mean serum concentrations of rIFN-αA as measured by bioassay for three patients treated on a twice daily schedule for 28 days with a total daily dose of 54 × 10^6 U to 27 × 10^6 U/injection every 12 hr. Within a few days of the initiation of therapy, a steady state level is reached which is maintained for the remainder of the study. Thirty-six hr after the final dose, however, serum levels of rIFN-αA are essentially undetectable.

Figure 5 shows the results of 9 patients treated three times weekly at 50 × 10^6 U/injection. Although the peak serum levels achieved are somewhat higher than on the twice daily study at a comparable dose per injection, no steady state serum level is achieved. In the 2- or 3-day intervals between injections, rIFN-αA levels become undetectable.

Patients treated with daily injections of rIFN-αA showed a tendency to achieve a steady state level.

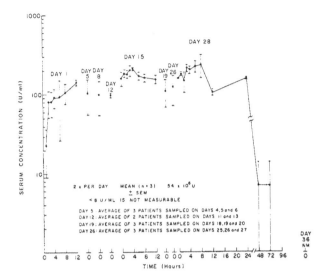

FIGURE 4. Serum concentration as measured by bioassay (I.M. ADMINISTRATION).

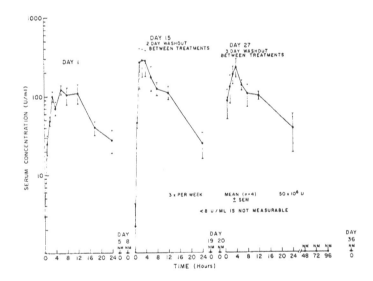

FIGURE 5. Serum concentration as measured by bioassay (I.M. ADMINISTRATION).

It should be emphasized that the importance of serum levels of interferon in terms of exerting a given biologic activity remains unknown. Interferon may achieve certain biologic effects indirectly through modulation of the immune system, in which case it could exert an effect in a body compartment into which it did not itself directly penetrate. The relationship of serum interferon levels and anti-tumor or antiviral activity will require a great deal more study.

D. Antibodies to rIFN-αA

The development of antibodies to rIFN-αA and HuIFNα was tested for by using a modification of the bioassay in which neutralization of the antiviral activity of interferon-containing samples was measured.

As shown in Table 3, of 201 different patients entered into these studies, 11 were found to have neutralizing IgG antibody to rIFN-αA. Two of these patients with the highest titers to rIFN-αA also had antibody to Cantell interferon. Studies at Hoffmann-La Roche with sera from 76 cancer patients treated solely with HuIFNα at M.D. Anderson, Memorial Sloan-Kettering Cancer Center, and Wadley Institute of Molecular Biology, showed 5 patients with antibody to rIFN-αA. Three of these developed antibody after treatment and two were found to have spontaneously occurring antibody at baseline prior to any treatment with exogenous interferon. In three of these patients, including the two with pre-existing antibody, neutralizing antibody to Cantell interferon was also detected.

This documents the occurrence of antibodies to naturally occurring endogenous interferons, at least in cancer patients. Of interest, is the apparent lack of any unusual susceptibility in these patients to viral infection in terms of frequency and severity. It is also likely that the occurrence of antibodies to Cantell interferon has been underestimated for a variety of reasons. In many studies, antibody testing was not performed at all. Further, the assay commonly used to detect interferon antibodies is a neutralization test of interferon's antiviral activity. If the target interferon is a mixture of the various leukocyte species, some of which are not antigenically cross-reactive, antibody which develops to a specific species may be missed. It is necessary to test for antibodies to each leukocyte species in addition to the natural mixture as represented by Cantell interferon in order to accurately assess the incidence of anti-interferon antibodies. No clinical or laboratory abnormalities were observed in the antibody positive patients.

E. Antitumor Activity of rIFN-αA

Table 4 illustrates tumor response observed at doses of rIFN-αA somewhat arbitrarily defined as low, intermediate, and high. The doses listed are per injection, not cumulative weekly doses. However, the size of the dose per injection per day correlates generally with the size of cumulative weekly doses on the three evaluated schedules.

rIFN-αA showed substantial activity against non-Hodgkin's lymphoma, mainly in the favorable histologies, at both high and low doses. Eight to 12 patients with the syndrome of Kaposi's sarcoma and acquired immunodeficiency responded at doses of 36 and 54 × 10^6 U daily with two complete responses (CR), three partial responses (PR), and three minimal responses (MR). Activity against multiple myeloma was observed but only at high doses, and the two partial responses occurred in patients with IgA myeloma and plasmocytoma. No significant responses in IgG myeloma were noted. Some activity was noted in renal cell cancer and melanoma. Somewhat surprisingly, only one partial response and two minimal responses were seen among 28 breast cancer patients treated, a much lower response rate than might have been anticipated on the basis of reports of HuIFNα. Possible explanations include differences in the biologic activities of various leukocyte interferon species and differences in patient populations, which in our trials were refractory to chemotherapy.

rIFN-αA showed little activity against colon and non-small cell lung cancer and soft tissue sarcomas other than Kaposi's sarcoma. Other malignancies are listed but were treated in numbers too small to comment upon. Statements about true response rates and dose response are not possible from these phase I studies because most patients in these studies were heavily pretreated and refractory, and dosage groups were sequentially entered and not necessarily comparable. Phase II disease specific pilot studies are now being performed.

F. Immunomodulating Effects of rIFN-αA

Results of the immunologic analysis of 134 patients treated with rIFN-αA on the three times weekly and twice daily studies were evaluated and included NK activity and monocyte mediated growth inhibition tests.[17] Each patient was evaluated 2 or 3 times prior to receiving rIFN-αA to establish accurate baselines. Subsequent samples were drawn on days 1, 4, 7,

Table 3

SERUM NEUTRALIZING ANTIBODY TITERS

Patient	Dose of rIFN-αA	rIFN-αA Pre	rIFN-αA Post	Cantell Post
SW	3—72 × 10⁶U	<100	600	<100
JK	3—108 × 10⁶U	<100	150	<100
	36 × 10⁶U, QD × 31 days	—	204,800	2,400
MJ	3—144 × 10⁶U	<100	400	<100
	36 × 10⁶U, QD × 28 days	500	1,400 (day 15)	<100
			280 (day 6≥)	<100 (day 6≥)
MO	36 × 10⁶U, QD × 28 days	<100	200	<100
	36 × 10⁶U, 3 × w × 28 days	1,600	265,000	300
FW	9 × 10⁶U, QD × 28 days	<100		<100
CP	18 × 10⁶U, QD × 28 days	<100	2400	<100
AWC	1.5 × 10⁶U, BID × 28 days	<100	300	<100
GR	9 × 10⁶U, BID × 28 days	<100	100	<100
CF	118 × 10⁶U, 3 × w × 12 wks.	<100	260	<100
RC	36 × 10⁶U, QD × 28 days	<100	<100	<100
	36 × 10⁶U, 3 × w × 12 weeks	<100	300	<100
RA	18 × 10⁶U, QD × 28 days	<100	<100	<100
	18 × 10⁶U, 3 × w × 28 days	<100	1000	<100

Note: Eleven of 201 patients developed neutralizing antibody to rIFN-αA. One of these patients with the highest titer also had antibody to Cantell Interferon.

19, 28, and 36. The NK cytotoxicity assay used K-562 target cells and effector target ratios of 60:1, 20:1 and 7:1 in triplicate. Monocyte function was assessed by the ability of the patient's cells to inhibit the growth of an NK-resistant tumor target cell line MB 12 in a 2-hr assay. Effector target ratios were 90:1, 30:1, 10:1, and 3:1 and growth inhibition activity was determined by the reduction in uptake of ^{3}H-thymidine by the target cells.

The percentage of post-treatment values of NK activity showing an increase, decrease, or no change from the limits of variability defined for that individual was evaluated for the twice daily and three times weekly studies. In the twice daily study, 6% of patients experienced an increase, 30% a decrease and 64% no change in NK activity. In the study evaluating 3 times weekly schedule, the results were 5%, 20%, and 75%, respectively. In the control group of normals not receiving interferon, 13% had an increase, 10% a decrease, and 77% no change of NK activity when evaluated over the same time span.

These data contrast with the results of in vitro testing of rIFN-αA on these patients' cells in which exposure to a wide range of interferon concentrations consistently enhanced NK activity.

To examine in more detail the depth of depression observed, it was determined whether the mean of the post-teatment values was significantly different from the pretreatment mean by a T-test, for low (<60 × 10⁶ U/week) intermediate (90 to 204 × 10⁶ U/week) and high (>250 × 10⁶ U/week) doses in the twice daily and three times weekly studies. Evidence of a dose response was present. There was an increase in the percentage of patients from 30% to 40% showing a significant decrease on the twice daily schedule and from 10% to 33% on the three times weekly schedule. The data suggest greater inhibition of NK activity for the more frequent schedule.

Table 4
ANTITUMOR ACTIVITY OF rIFN-αA— PHASE I STUDIES

	Low (0.1,1,3,9 × 10^6 U)	Intermediate (18 × 10^6 U)	High (≥36 × 10^6 U)
Breast	9 (1 PR, 1 MR)	4 (0)	15 (1 MR)
NHL	11 (3 PR, 3 MR)	3 (1 PR, 1 MR)	11 (3 PR, 3 MR)
Myeloma	2 (0)	0	6 (2 PR, 2 MR)
Kaposi's sarcoma	0	0	12 (2 CR, 3 PR, 3 MR)
Melanoma	4 (1 CR)	2 (0)	11 (1 PR)
Renal	3 (0)	2 (1 MR)	2 (1 PR)
Colon	9 (0)	5 (0)	22 (2 MR)
Lung (NSM)	5 (0)	0	4 (0)
Soft tissue sarcoma	6 (1 MR)	1 (0)	5 (0)
Hodgkin's disease	1 (1 MR)		1 (0)
Prostate	1 (0)	1 (0)	2 (1 MR)
Hepatoma	0	1 (1 MR)	3 (0)
Ovarian	1 (0)	1 (0)	3 (0)
Mesothelioma	3 (1 MR)	0	0

a dose response was present. There was an increase in the percentage of patients from 30% to 40% showing a significant decrease on the twice daily schedule and from 10% to 33% on the three times weekly schedule. The data suggest greater inhibition of NK activity for the more frequent schedule.

Monocyte function, however, showed a significant increase after the start of rIFN-αA therapy in both studies, with approximately 80% increasing, 3% decreasing, and 17% showing no change. The control group of normals not receiving interferon showed 7% increase, 9% decrease, and 84% no change. The level of activity in the interferon patients remained high for the duration of treatment and returned to pretreatment levels after therapy was stopped on day 28. No dose response was observed.

IV. PHASE I CLINICAL STUDIES OF rIFN-αA IN PATIENTS WITH VIRAL INFECTIONS

Because the protein content of cell culture derived Cantell interferon was composed mostly of proteins which were not interferon, it was hoped that the side effects observed with parenteral administration of that mixture would be due to substances other than interferon. As we have shown in our review of the tolerance studies run with pure, recombinant DNA derived interferon, the "flu-like" and other side effects appear to be due to interferon itself. This finding limits the use of parental interferon in viral infections to mostly severe diseases.

A. Parenteral Administration of rIFN-αA
1. Tolerance in Bone Marrow Transplant Patients
A problem which might be expected in administering interferon parenterally to patients with severe viral infections is that those who are most in need of the therapy are also at the most risk from the side effects. Bone marrow transplant recipients are at risk of developing fatal viral infections most particularly by cytomegalovirus (CMV). A significant concern in administering interferon parenterally to these patients is the fate of their graft. Studies evaluating rIFN-αA and conducted at UCLA and the Fred Hutchinson Cancer Center are reassuring in regard to this concern.[18,19] Bone marrow transplant recipients diagnosed clinically to have life-threatening viral infections were given rIFN-αA intramuscularly daily at doses of up to 50 × 10^6 units per day. There was no consistent effect on total white blood

cell counts or neutrophil counts. The remaining rIFN-αA clinical side effects were similar to those observed in cancer patients. There were too few patients to make a comment on efficacy.

2. Tolerance and Antiviral Effects in Chronic Active Hepatitis B Patients

Chronic hepatitis B is a potentially fatal disease. Previous work done at Stanford has shown that when patients with chronic hepatitis B were given Cantell interferon, they experienced decreases in their serum concentration of hepatitis B virus (HBV)-dependent DNA polymerase (DNAP).[20] Similar results were found using rIFN-αA.[21] The magnitude of decrease of DNAP, a parameter related to Dane particle concentration, was found to be related to the magnitude of the dose of rIFN-αA administered. The patients in this study received doses up to 68×10^6 U/day and all clinical and laboratory side effects noticed were similar to those observed in cancer patients.

B. Intranasal Administration of rIFN-αA

Except under special circumstances the viral infections which affect the lives of most people are annoying rather than life-threatening. Although the parenteral use of interferon in these diseases is probably not indicated, some of these diseases may be responsive to the topical use of interferon. The common cold is an example of such a disease. A rhinovirus challenge study using Cantell interferon sprayed into the noses of subjects 3 times in 1 hr, 4 times a day, showed efficacy in preventing colds.[22] A study giving rIFN-α2 by the same schedule also prevented colds from developing.[23] A different regimen was utilized in a rhinovirus and coronovirus challenge study using rIFN-αA in which subjects given two sprays of rIFN-αA per nostril twice a day, were protected from developing colds.[24] Although no significant adverse experiences occurred in these short-term duration intranasal studies, the route may not be an entirely benign one for the administration of interferon. Nasal congestion and epistaxis has been reported with intranasal administration of buffy coat derived interferon.[25] In this setting it would be reasonable to delay exposure of large numbers of volunteers to intranasal interferon until the risk of adverse reactions is better defined by tightly controlled tolerance studies using small numbers of volunteers.

V. SUMMARY AND CONCLUSIONS

The phase I studies described above represent the most systematic and extensive clinical evaluation of any interferon to date. It has been clearly shown that purified recombinant leukocyte interferon-rIFN-αA can be safely given to patients with various malignancies and viral diseases and that it has potent biologic activities in vivo. The clinical and laboratory side effects of rIFN-αA are similar to those observed with partially purified natural leukocyte interferons, manifest some degree of dose relationship, and strongly suggest that these toxicities represent intrinsic properties of interferon itself. While interferons appear relatively safe, even purified recombinant interferon is not well tolerated when administered parenterally except at the lowest doses tested. Poor tolerance may limit its utility in certain chronic and non-life threatening conditions.

Fatigue is the most frequent dose limiting side effect for chronic administration of 1 month or longer. Partial tolerance to fever and flu-like symptoms commonly develops and these symptoms do not generally present an obstacle to long-term treatment. Laboratory abnormalities are usually not dose limiting even in patients with disseminated cancer and compromised bone marrow reserve and liver function. Antitumor activity in vivo particularly against favorable non-Hodgkin's lymphoma and Kaposi's Sarcoma was observed, as was antiviral activity against chronic active hepatitis B and prevention of some common cold viruses following intranasal administration.

Immunomodulatory effects of rIFN-αA are variable and seem not to be dose- or schedule-specific. We were not yet able to correlate changes of immune parameters with other biologic activities or to define the optimal immunomodulatory dose. We need now to define the true response rates in various cancers and viral diseases with optimal regimens of rIFN-αA. In the absence of immunomodulatory dose, empirical comparisons of high and low doses and various schedules of rIFN-αA may be necessary. In addition, several other leukocyte interferon species, hybrids of species, fibroblast, and immune interferons have been cloned. Extensive preclinical testing should be done to help determine the most promising single interferon as well as the effects of combinations of them in malignant and viral diseases. Eventually, evaluation of recombinant interferon in combination with other anticancer and antiviral agents will likely be indicated. The role of parenterally and topically administered recombinant interferons in the adjuvant setting for cancer and as prophylaxis for certain viral disease should be evaluated. This may prove to be the most important clinical use of interferon. Phase II efficacy trials of rIFN-αA are presently underway in a variety of cancers and viral diseases to answer some of the many questions which remain.

REFERENCES

1. **Isaacs, A. and Lindemann, J.**, Virus interference. I. The interferons. *Proc. R. Soc. London Biol.*, 147, 258, 1957.
2. **Paucker, K., Cantell, K., and Henle, W.**, Quantitative studies on viral interference in suspended L-cells III. Effect of interfering viruses and interferon on the growth rate of cells, *Virology*, 17, 324, 1962.
3. **Gresser, I., Bourali, C., and Levy, J. P.**, Increased survival in mice inoculated with tumor cells and treated with interferon preparations. *Proc. Natl. Acad. Sci. U.S.A.*, 63, 51, 1969.
4. **Gresser, I., Broum-Boye, D., Thomas, M. T., and Maciera-Coelho, A.**, Interferon and cell division I. Inhibition of the multiplication of mouse leukemia lizio cells in vitro by interferon preparations, *Proc. Natl. Acad. Sci. U.S.A.*, 66, 1052, 1970.
5. **Atanasiu, P. and Chany, C.**, Action d'un interferon provenat de Cellules malignes sur l'infection experimentale du hamster nouveau-ne par le virus du polyome, *C.R. Acad. Sci. (Paris)*, 251, 1687, 1960.
6. **Merigan, T. C., Rand, K. H., Pollard, R. B., Abdallah, P. S., Jordan, G. W., and Fried, R. P.**, Human leukocyte interferon for the treatment of herpes zoster in patients with cancer, *N. Engl. J. Med.*, 298, 981, 1978.
7. **Cheeseman, S. H., Rubin, R. H., Stewart, J. A., Tolkoff-Rubin, N., Cosini, A. B., Cantell, K., Gilberg, J., Winkle, S., Herrin, J. F., Black, P. H., Russell, P. S., and Hirsch, M. S.**, Controlled clinical trial of prophylactic human leukocyte interferon in renal transplantation, *N. Engl. J. Med.*, 300, 1345, 1979.
8. **Dziewanowska, Z. E. and Pestka, S.**, The human interferons, *Medicinal Research Reviews*, 2, 4, 325, 1982.
9. **Goeddel, D. V., Yelverton, E., Ullrich, A., Heynecker, H. L., Miozzari, G., Holmes, W., Seeburg, P. H., Dull, T., May, L., Stebbing, N., Crea, R., Maeda, S., McCandliss, R., Sloma, A., Tabor, J. M., Gross, M., Familletti, P. C., and Pestka, S.**, Construction and identification of bacterial plasmids containing nucleotide sequence for human leukocyte interferon, *Proc. Natl. Acad. Sci. U.S.A.*, 3, 77, 7010, 1980.
10. **Ortaldo, J., Herberman, R., Mantovani, A., Hobbs, D., Kung, H., and Pestka, S.**, Effects of recombinant interferon on sideotoxic activity of natural killer (NK) cells and monocytes, *Cell. Immunol.*, 67, 160, 1982.
11. **Veroma, D. S., Spitzer, G., Gutterman, J. U., Johnston, D. A., McCredie, K. B., Hobbs, D. S., Kung, H.-F., and Pestka, S.**, in preparation.
12. Data on File, Hoffmann-La Roche Inc.

13. **Gutterman, J. U., Fein, S. H., Quesada, J., Horning, S. J., Levine, J. L., Alexanian, R., Bernhardt, L., Kramer, M., Speigel, H., Colburn, W., Trown, P., Merigan, T., and Dziewanowska, Z.,** Recombinant leukocyte A interferon: Pharmacokinetics, single dose tolerance and biologic effects in cancer patients, *Ann. Int. Med.*, 5, 549, 1982.

14. **Sherwin, S. A., Knost, J. A., Fein, S., Abrams, P., Foon, K. A., Ochs, J. J., Schoenberger, C., and Maluish, A. E.,** A multiple dose phase I trial of recombinant leukocyte A interferon using a 3 × weekly schedule. American Soc. Clin. Onc., 18th Ann. Meeting, April 1982.

15. **Leavitt, R. D., Duffey, P. L., Wiernik, P. U., Fein. S. H., Sherwin, S., Scogna, D., and Oldham, R.,** A phase I trial of twice daily recombinant human leukocyte A interferon —IFL-rA— in cancer patients, Am. Soc. Clin. Onc., 18th Ann. Meeting, April 1982.

16. **Krown, S., Cunningham-Rundles, S., Fein, S. H., Koziner, B., Myskowski, P., Real, F., Safai, B., and Oettgen, H.,** Treatment of Kaposi's sarcoma and acquired immunodeficiency with recombinant leukocyte A interferon, in preparation.

17. **Maluish, A. E., Ortaldo, J. R., Sherwin, S., Leavitt, R. D., Strong, D., Fein, S. H., Wiernik, P., Oldham, R. K., and Herberman, R. B.,** Immunologic monitoring of patients receiving recombinant leukocyte A interferon, Am. Soc. Clin. Onc., 18th Ann. Meeting, April 1982.

18. **Winston, D. J., Ho, W. G., Schroff, R. W., Bartoni, K., Chomplin, R. E., and Gale, R. P.,** Initial studies of safety and tolerance of recombinant leukocyte A interferon (Ro 22-8181) in bone marrow transplant recipients. 22nd Intersci. Conf. *Antimicrob. Agents Chemother.* (ICAAC), October 1982, Abstract #191. 23, 846, 1983.

19. **Meyers, J. D., Day, L. M., and Lum, L. G.,** Safety and efficacy of recombinant leukocyte A interferon (IFL-rA) for the treatment of serious virus interferon after marrow transplant. 22nd Intersci. Conf. Antimicrob. Agents and Chemother., (ICAAC), October 1982, Abstract #189.

20. **Greenberg, S. B., Pollard, R. B., Lutwick, L. I., Gregory, P. B., Roberson, W. S., and Merigan, T. C.,** Efficacy of human leukocyte interferon on hepatitis B virus infections in patients with chronic active hepatitis, *N. Engl. J. Med.*, 295, 517, 1976.

21. Data on File, Hoffmann-La Roche Inc.

22. **Merigan, T. C., Reed, S. E., Hall, T. S., and Tyrrell, D. A. J.,** Inhibitions of respiratory virus infection by locally applied interferon., *Lancet*, 1, 563, 1973.

23. **Scott, G. M., Wallace, J., Greiner, J., Phillpotts, R. J., Gauci, C. L., and Tyrrell, D. A. J.,** Prevention of rhinovirus colds by human interferon alpha-2 from *Escherichia Coli, Lancet*, 2, 186, 1982.

24. **Samo, T. C., Greenberg, S. B., Couch, R. B., and Quarles, H. J.,** Protection against illness by recombinant leukocyte A interferon in rhinovirus challenged volunteers. Abstract submitted to Southern Society for Clinical Investigation. September, 1982, *J. Infect. Dis.*, 148, 535, 1983.

25. **Scott, G. M., Phillpotts, R. J., Wallace, J., Secher, D. S., Cantell, K., and Tyrrell, D. A. J.,** Purified interferon as Protection against rhinovirus infection, *Br. Med. J.*, 284, 1822, 1982.

Chapter 7

LARGE SCALE PRODUCTION OF HUMAN ALPHA INTERFERON FROM BACTERIA

W. Courtney McGregor and Armin H. Ramel

TABLE OF CONTENTS

I. INTRODUCTION

In considering the production of interferons, we must keep in mind that scale of operation is relative. The low physiological levels of interferon found in nature suggest that it is relatively potent at low dose levels and that the amounts needed for clinical use may be very small compared to other drugs. In recent years, a few milligrams of interferon produced by classical tissue culture methods were considered a large quantity and optimistic projections for future commercial production by recombinant technology refer to kilograms. This is in sharp contrast to antibiotics, for example, where worldwide production is in the order of thousands of tons. Thus, production of interferon may be considered large to a biochemist, but may be perceived as not more than pilot plant scale to a chemical engineer involved in conventional pharmaceutical manufacturing.

Production can be conveniently broken down into two major components: fermentation and recovery. The scale of operation of both components will depend greatly on the expression levels of interferon in the bacteria. It has been reported that yields of recombinant protein in bacteria may be very substantial, as high as 30 to 50% of the soluble protein synthesized in the cell.[1] Let us assume for our purposes that only 1% of the total cell protein is recombinant interferon, that it stays in the cell, and that 10% of the wet weight of the cell is protein. Every kg of cells would produce 1 g of interferon. If we further assume a 10% recovery during purification, then we would need 10 kg cells to produce 1 g interferon. To keep the numbers simple, let us assume an annual requirement of 10 kg interferon and that the fermentor produces 30 g cells/ℓ. We would need to process 100,000 kg of cells per year or about 400 kg per day. Daily fermentation requirements would be roughly 15,000 ℓ.

In the fermentation industry, this is a relatively small order. On the other hand, there is little precedent for this scale of protein purification. We face a challenge not only to produce relatively large quantities of a biolymer for human therapy, but also to provide it at high purity. It is expected that improvements in purification techniques as well as improvements in analytical techniques to monitor purity will drive the requirement for purity ever higher.

This chapter will consider some practical aspects of production in the three areas mentioned above: fermentation, recovery, and purity analysis.

II. FERMENTATION

As described in previous chapters, interferon can be cloned and expressed in yeast and mammalian cells as well as bacteria. Thus far, however, all recombinant interferon for clinical use has been produced in the Gram-Negative bacteria, *Escherichia coli (E. coli)*. Reasons for selecting *E. coli* are fairly straightforward. The whole evolution of molecular biology has been intimately connected with the study of *E. coli*, its plasmids, and its phages. The genetics and physiology are relatively simple and well characterized. Fermentation growth rate is rapid.

An advantage that other organisms such as yeast or bacillus may have is the capability of secreting the product into the fermentation broth. It has been speculated that this would make the subsequent purification steps easier.

Prior to the advent of recombinant DNA technology, *E. coli* had been used in the fermentation industry primarily for the production of amino acids.[2] The only clinically important fermentation product from *E. coli* was L-asparaginase, an enzyme used in the treatment of cancer.[3] The point here is that industrial scale fermentation of *E. coli* exists as an established practice and presents no major technical hurdle per se.

There is, however, a significant feature of the NIH Guidelines for Recombinant DNA Research which applies to large-scale fermentation. When cells are harvested from fermentors larger than 10 ℓ, they must be killed before leaving the fermentor or, alternatively, the broth may be harvested in 10 ℓ aliquots. In the latter case no live recombinant cells must be allowed to escape into the environment. Procedures must be followed which provide for the killing of all cells including cellular debris before disposal.

These requirements while not all that formidable have an important effect on the downstream side of the process. In selecting a killing process for the fermentor, one must consider the effects not only on the biological activity of the interferon but also on the subsequent purification process itself. A variety of approaches are possible and include heat, phenol, hypochlorite, acid/base, and peroxide.

What about the pyrogenic lipopolysaccharide or endotoxin from the *E. coli* cell envelope ending up in the product? This question is raised by many investigators, especially because endotoxin elicits a pyrogenic response at very low concentrations, i.e., 0.5 to 1.0 ng/kg of body weight. On the other hand, lipopolysaccharides would not be expected to co-purify with interferon. First, lipopolysaccharides are high molecular weight generally of the order of millions. Second, they are non-protein and non-ionic. In the usual labyrinth of purification steps these kinds of contaminants would be expected to be removed provided the water is pyrogen-free to begin with.

Fermentations for production of human leukocyte interferon have been carried out at Hoffmann-La Roche according to established procedures and killed before harvesting during late logarithmic growth. The cell pellets collected by centrifugation have been stored at −20°C until used.

III. RECOVERY

Large-scale recovery of product from fermentation cell mass can be conveniently subdivided into two parts: isolation and purification. The isolation steps are usually relatively large in volume beginning with disruption of the resuspended cells by chemical or mechanical means. Cell debris removal, typically by centrifugation, is followed by one or more volume reduction steps. Precipitation with salts such as ammonium sulfate is a common method for reducing volume and it simultaneously provides a modest degree of purification. From this point the process can be classified as purification and usually includes several types of chromatography columns. The classical gel filtration and ion exchange columns used for the last 20 years in protein purification have been supplemented more recently with newer adsorption media. These include a variety of affinity adsorbents with a high degree of specificity. Enzyme inhibitors, substrate analogues, organic dyes, metal chelates, and antibodies are being used as ligands for affinity purifications.

In our goal to approach 100% homogeneity we will be restricted by the yield/purity trade-off. With each additional chromatography column to achieve greater purity there will be an unavoidable loss of product. It is, therefore, very attractive to adopt one of the most powerful types of affinity separations such as immunoadsorbent chromatography. With immobilized monoclonal antibodies, which are highly selective for single proteins, it is conceptually possible to approach, if not actually accomplish, the so-called single step purification. Although the technique for using immobilized monoclonal antibodies for purification of proteins is relatively simple, the preparation of the antibodies requires a major effort.[4]

One of the current methods of producing recombinant human leukocyte interferon takes advantage of the above mentioned developments in protein purification. A report on the use of monoclonal antibodies indicates an overall recovery of interferon up to 81%.[5] The promise of repeated use of monoclonal antibodies in affinity columns makes this approach attractive for the preparation of large amounts of clinical material.

IV. PURITY AND IDENTITY

Many readers may not appreciate the variety of regulatory laws which must be complied with in the manufacture of drugs including recombinant interferon. The laws are formulated in the Food, Drug, and Cosmetic Act, the Public Health Service Act, and the Radiation Control for Health and Safety Act. Detailed requirements for compliance can be found codified in the Code of Federal Regulations. In addition, all manufacturers of drugs must operate in compliance with current Good Manufacturing Practices. All these regulations enforced by the Food and Drug Administration are meant to ensure that the public receives only drugs whose identity, purity, safety, and efficacy have been demonstrated. Those of us involved in production must be concerned with identity and purity. Others in the organization may then test for safety and efficacy.

The identity of recombinant interferon can be established not only by the antiviral or cytopathic effect inhibition assay[6] but also at the genomic level. For example, the successful construction of the plasmid used for the transformation of the *E. coli* can be periodically verified by DNA sequencing. Other identity tests include total amino acid analysis of the product and the classic methods of N- and C-terminal amino acid analysis in conjunction with proteolytic peptide mapping. Recent developments in separating proteins by reverse-phase high performance liquid chromatography (HPLC) columns now permit analysis of subnanomole quantities of complex peptide mixtures.[7] Isoelectric focusing, radio- or enzyme immunoassays, and ultimately amino acid sequencing are additional methods. Taken together these methods permit the unambiguous identification of human alpha interferon produced in bacteria.

Purity on the other hand is a relative term. It is defined by the assay used to monitor purity and is bound by the technical limits of that assay. For example, the conventional method for detecting protein impurities is electrophoresis in polyacrylamide gel in presence of the detergent sodium dodecyl sulfate and the reducing agent mercaptoethanol. Protein bands separate on the basis of molecular weight and can be detected by staining with Coomassie Blue which has a detection limit of about 0.1 μg protein or 1% of the 10 μg load. Newer silver stains have been reported to be up to 100 times more sensitive.[8]

Reverse-phase or size exclusion HPLC can also be used for monitoring protein impurities. HPLC has the advantage of being faster than electrophoresis, although sensitivity comparisons with electrophoresis are not available.

The high degree of purity of recombinant leukocyte interferon in clinical testing (above 95%) is made possible by the use of genetic engineering. From about 12 subtypes of human leukocyte interferons found in conventional tissue culture preparations, only one is produced in a recombinant *E. coli* fermentation. Furthermore, the new developments in affinity purifications mentioned above have afforded improvements in specific activity up to 100-fold. For example, the specific activity of human alpha interferon produced by conventional tissue culture and conventional purification techniques is reported to be 2.5×10^6 to 1.25×10^7 IU/mg.[9,10] On the other hand, recombinant leukocyte interferon A purified by monoclonal antibodies has been reported as high as 3×10^8 IU/mg.[5]

It is important to emphasize that impurities other than protein must also be monitored depending on the production process and the ultimate use of the drug. Endotoxins or pyrogens, for example, would not be detected in a protein assay but if the interferon is to be formulated as a parenteral drug, pyrogens must be below a certain standard in the rabbit injection assay.

A combination of the technologies described above have made it possible for Hoffmann-La Roche to sustain the clinical trials begun on January 15, 1981 with gram quantities of highly pure human leukocyte interferon.

REFERENCES

1. **Miozzari, G. F.**, Strategies for obtaining expression of peptide hormones in *E. coli*, in *Insulins, Growth Hormone and Recombinant DNA Technology*, Gueriguian, J. L., Ed., Raven Press, New York, 1981.

2. **Hirose, Y. and Okada, H.**, Microbial production of amino acids, in *Microbial Technology*, Peppler, H. J. and Perlman, D., Eds., Academic Press, New York, 1979.

3. **Grinnan, E. L.**, L-Asparaginase: A case study of an *E. coli* fermentation product, in *Insulins, Growth Hormone and Recombinant DNA Technology*, Gueriguian, J. L., Ed., Raven Press, New York, 1981.

4. **Staehelin, T., Durrer, B., Schmidt, J., Takacs, B., Stocker, J., Miggiano, V., Stahli, C., Rubinstein, M., Levy, W. P., Hershberg, R., and Pestka, S.**, Production of hybridomas secreting monoclonal antibodies to the human leukocyte interferons, *Proc. Natl. Acad. Sci.*, 78, 1848, 1981.

5. **Staehelin, T., Hobbs, D. S., Kung, H. F., Lai, C.-Y., and Pestka, S.**, Purification and characterization of recombinant human leukocyte interferon (IFLrA) with monoclonal antibodies, *J. Biol. Chem.*, 256, 9750, 1981.

6. **Rubinstein, S., Familletti, P. C., and Pestka, S.**, Convenient assay for interferons, *J. Virol.*, 37, 755, 1981.

7. **Rubinstein, M., Stein, S., and Udenfriend, S.**, Flurometric methods for analysis of proteins and peptides; principles and application, *Horm. Proteins Peptides*, 9, 1, 1980.

8. **Wray, W., Boulikas, T., Wray, V. P., and Hancock, R.**, Silver staining of proteins in polyacrylamide gels, *Anal. Biochem.*, 118, 197, 1981.

9. **Horowitz, B.**, Human interferon — properties, clinical application, and production, *J. Parenteral Sci. Technol.*, 35, 223, 1981.

10. **Cantell, K., Hirvonen, S., and Koistinen, V.**, Partial purification of human leukocyte interferon on a large scale, *Meth. Enzymol.*, 78, 499, 1981.

Chapter 8

DIRECT EXPRESSION OF HUMAN GROWTH IN *ESCHERICHIA COLI* WITH THE LIPOPROTEIN PROMOTER

Nancy G. Mayne, Hansen M. Hsiung, John D. Baxter, and Rama M. Belagaje

TABLE OF CONTENTS

I. SUMMARY

A plasmid for the expression of the human growth hormone (hGH) gene was constructed using a novel combination of a cloning vector, pKEN024, containing the lipoprotein gene of *Escherichia coli*, ptrpED50chGH (containing cDNA to hGH mRNA), and a chemically synthesized DNA fragment that linked the hGH cDNA to the lipoprotein gene promoter and ribosome-binding site. The hybrid gene was expressed in *E. coli*, and its product accounted for about 10% of the bacterial protein synthesized. This product, methionyl hGH, was isolated and its biological activity was found to be similar to that of natural hGH obtained from cadavers. This procedure provides a means to obtain large quantities of biologically active hGH.

II. INTRODUCTION

Human growth hormone (hGH) comprises a single polypeptide chain of 191 amino acids with a molecular weight of 22,000. This hormone is limiting for linear growth in man and it also has a number of other biological activities.[1-5] The current use of hGH in therapy is restricted to the treatment of dwarfism due to growth hormone deficiency.[1-3] However, this hormone may also be useful for the treatment of certain types of growth retardation not known to be due to growth hormone deficiency[4] and for a variety of conditions that are unrelated to growth deficiency.[4] Since non-primate growth hormones have not been found to have significant biological activity in man, the availability of growth hormone for use in humans has been restricted to that hGH obtained from cadavers. This has therefore limited the availability of the hormone, which cannot be synthesized in high quantity by conventional techniques, for experimental testing of its utility in treating other disorders.

Recent advances in recombinant DNA technology have provided the means to produce proteins in bacteria that have not been obtainable in large quantity by other techniques. Thus, to date, sequences coding for several mammalian proteins have been cloned and expressed in bacteria.[6-15] In this respect, we earlier cloned in bacteria and sequenced cDNA to hGH and constructed a plasmid that directed the synthesis of a protein that contained hGH bound to amino acids of the tryptophan operon gene product.[11] Following this, Goeddel and coworkers, utilizing this cDNA and synthetic DNA fragments, reported the expression of methionyl hGH under the control of a double promoter for bacterial β-galactosidase.[14] In the current report, we describe methodology for the synthesis of methionyl hGH with the use of the promoter for the outer-membrane lipoprotein of gram-negative bacteria,[16-23] hGH cDNA and a synthetic cDNA fragment. The lipoprotein (*lpp*) gene codes for the most abundant protein in bacteria and its nucleic acid sequence has been determined.[23,24] Inouye et al.[16-23] have studied this gene extensively and have constructed both constitutive and inducible expression-cloning vehicles.[24,25,40] In the present study we have used the PIN-I-type constitutive vehicle. The results show that methionyl hGH can be synthesized in bacteria with the use of this technique and that this synthesis can comprise 10% of the total bacterial protein synthesis.

III. MATERIALS AND METHODS

Restriction endonucleases and DNA polymerase I (large fragment) were obtained from New England Biolabs, bacteriophage T_4 DNA ligase from Bethesda Research Laboratories, bacteriophase T_4 polynucleotide kinase from Collaborative Research, calf intestine alkaline phosphatase from Boehringer Mannheim, and S1 nuclease from Miles Laboratories. {γ-^{32}P}ATP was purchased from Amersham. All enzymes were used in reaction mixtures recommended by the vendors.

Transformation of *E.coli* K-12 strains JA221 (hsdM⁺hsdR⁻leuB6ΔtrpE5) and HB101 (Pro leu thi lacY hsdM⁻hsdR⁻endA rpsL20 ara14 galK2 xy15 mt1-1 supE44) was performed by a modification of the method of Wensink et al.[26] Plasmids pKEN024[41] and ptrp-ED50chGH[11] were isolated by the alkaline miniscreen procedure and examined by restriction enzyme analysis. For sequencing by the Maxam and Gilbert procedure,[27] plasmids were prepared by CsCl-ethidium bromide centrifugation of cleared cell lysates.[28]

Bacterial cultures containing plasmids were grown to an OD_{590} of one in brain heart infusion broth (Gibco®). Pellets obtained from 1 mℓ of these cells were lysed and separated on a 12.5% SDS polyacrylamide gel. For radioimmunoassays, similar cell pellets were lysed with lysozyme and 0.1% Triton X-100®. These whole cell preparations were assayed for the presence of hGH using hormone standards and antibodies obtained from Cambridge Nuclear.

IV. RESULTS

To construct a plasmid to direct the synthesis of methionyl hGH, we used the lipoprotein gene promoter and ribosome-binding sites contained within pKEN024 (Figure 1), the previously cloned hGH cDNA in ptrpED50chGH800 (Figure 1), a synthetic DNA fragment and sequences from pBR322 (Figure 2). Steps in the construction are outlined in Figures 1 to 3.

The structure of the chemically synthesized double-stranded DNA fragment used to join the lpp promoter and ribosome-binding site to the hGH coding region is shown in Figure 2. The upper strand has 66 nucleotides which include, on the 5′-end, the 4 nucleotide single-stranded sequences produced by Xba I cleavage. The lower strand has 62 nucleotides. The synthetic DNA contains the natural sequence of the lpp gene from the Xba I site in the ribosome-binding site (16 bp), the translation initiating methionine codon and the first 47 nucleotides (amino acid codons 1 to 15 and part of 16) of hGH cDNA ending with the unique Fnu DII site previously described.[11] In preparing this double-stranded DNA segment, the 11 deoxyoligonucleotides indicated by arrows (Figure 2) were all synthesized by the phosphotriester method as described previously with some modifications.[32,33] These oligo-nucleotide segments were then assembled to form the desired DNA fragment by performing four separate T_4 ligase catalyzed reactions as described in Figure 2. The authenticity of this structure was verified by a subsequent sequence analysis of the cloned fragment.

To prepare an hGH cDNA fragment suitable for connection to the synthetic DNA and to the plasmid containing the lpp promoter, ptrpED50chGH800 was cleaved at the unique Sma I site (Figure 1) and Bam HI linkers were added. The resulting fragment was cleaved at the unique Bam HI site (Figure 1) and the 691 bp fragment gel-isolated. This fragment was then inserted into the Bam HI site of pBR322. After amplification of this plasmid, the 691-bp fragment was re-isolated and cleaved with Fnu DII 47 nucleotides from the first amino acid between the codon for amino acid 16 of hGH cDNA, resulting in a 538-bp fragment with Fnu DII and Bam HI termini, the codons for amino acids 16 to 191 of hGH, the UAG stop codon and nine nucleotides of 3′-noncoding DNA.

To form the hGH expression plasmid, the 538-bp fragment described above and the synthetic DNA fragment were ligated to pKEN024, which had been treated with Bam HI and Xba I and, subsequently, alkaline phosphatase (Figure 1), forming pNM645. The Xba I site is located in the ribosome-binding site 16 bp before the initiation codon, whereas the Bam HI site is located in the coding sequence for the lpp gene, and cleavage at this site leaves the codons for the last 35 amino acids of lipoprotein remaining in the fragment to be cloned. This fragment was also modified as shown in Figure 3 to add the tetracycline-resistant gene from pBR322. This was achieved by replacing the lpp 3′-sequence between the Bam HI and Sal I restriction sites with a DNA fragment derived from pBR322.

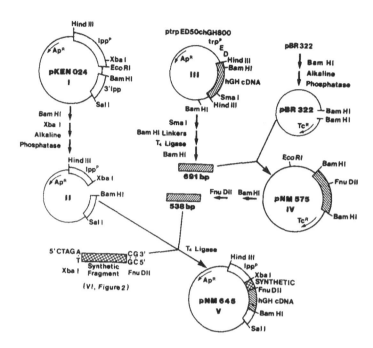

FIGURE 1.   Construction of a plasmid (pNM645) for the bacterial expression of human growth hormone. Ten micrograms of pKEN024 (I), with a 462-bp fragment containing the promoter, 5'-untranslated region and ribosome binding site of the *E. coli lpp* gene[24,25,40] were digested in 200 μℓ of Xba I/Bam HI buffer using 10 units of Bam HI for 1 hr at 37°C followed by 10 units of Xba I for 1 hr at 37°C. The DNA was then treated with 2.5 units of alkaline phosphatase for 1.5 hr at 65°C, phenol/CHCl₃-extracted and collected by ethanol precipitation. This preparation (II) was used as the plasmid cloning vector.

ptrpED50chGH (III), described by Martial et al.,[11] was used as the source of a DNA fragment containing the coding sequences for amino acids 16 to 191 of hGH. The unique Sma I restriction site 6 bp downstream from the translation termination codon of the gene was changed to a Bam HI site by Sma I cleavage and bacteriophage T₄ DNA ligase addition of Bam HI linkers. Treatment of the DNA with Bam HI enzyme cleaved the attached linker sequence as well as a Bam HI site located at the beginning of the cloned cDNA sequence of hGH. This yielded a 691-bp fragment with cohesive Bam HI termini, which was separated on a 6% polyacrylamide gel and visualized under long-wave-length UV light after staining in an ethidium bromide solution at 1 μg/mℓ. The gel region containing the fragment was excised, and the DNA fragment was recovered by electroelution followed by ethanol precipitation. The recovered DNA fragment was then ligated with 0.2 μg of pBR322, which had been cleaved at its unique Bam HI site and treated with alkaline phosphatase. After 16 hr at 4°C, the material was used to transform *E. coli* strain JA221.[26] Transformed colonies were selected on agar plates containing 100 μg/mℓ of ampicillin. Plasmid DNAs were isolated from 16 of the ampicillin-resistant colonies by the rapid alkaline-denaturation method of Birnboim and Doly[29] and then analyzed by restriction enzyme digestion and gel electrophoresis. Eleven of the 16 plasmids examined were found to contain a Bam HI fragment of approximately 700 bp. One of these plasmids, pNM575 (IV), was amplified. Twenty-five micrograms of pNM575 were digested with Bam HI. The resulting 691-bp fragment was isolated from a 6% polyacrylamide gel, purified as described above, and digested with Fnu DII to yield a 538-bp DNA fragment containing the coding sequences for the last 175 amino acids of hGH followed by a translation stop codon. This fragment was isolated by electrophoresis on a 6% polyacrylamide gel.

The expression plasmid pNM645 was constructed by enzymatically joining 0.1 pmol (0.4 μg) of plasmid vector (II), 3.2 pmol of the synthetic DNA fragment (VI in Figure 2), and 0.24 pmol (0.08 μg) of the 538-bp fragment in 14 μℓ of ligation buffer using 2 units of bacteriophage T₄ DNA ligase. After incubation for 16 hr at 4°C, the mixture was used to transform *E. coli* JA221 as previously described. Transformed colonies were selected on agar plates containing 100 μg/mℓ ampicillin. Plasmids from 10 colonies were prepared by the previously described Birnboim screening procedure.[29] After digestion with Xba I and Bam HI, followed by acrylamide gel electrophoresis, one plasmid was found to contain the expected 604-bp fragment. This plasmid was amplified and the DNA sequence from the Xba I site through the Fnu DII site was determined by the Maxam and Gilbert method[27] and found to be correct.

The following abbreviations are used: Ap^r, ampicillin-resistant; Tc^r, tetracycline-resistant; and Tc^s, tetracycline-sensitive; . . . . . . *lpp* sequences; | | | | | synthetic DNA; \ \ \ \ \ hGH cDNA.

139

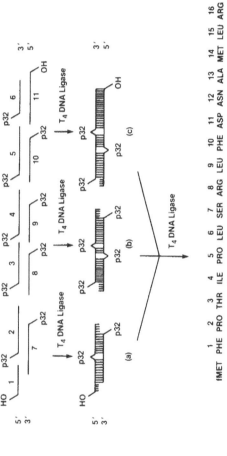

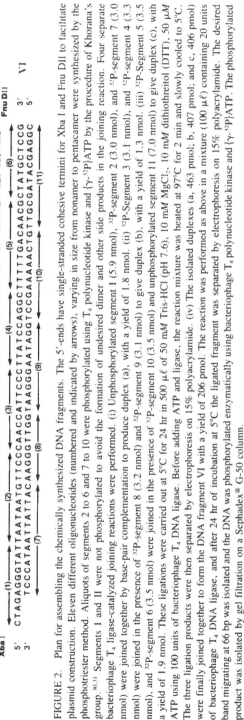

FIGURE 2. Plan for assembling the chemically synthesized DNA fragments. The 5'-ends have single-stranded cohesive termini for Xba I and Fnu DII to facilitate plasmid construction. Eleven different oligonucleotides (numbered and indicated by arrows), varying in size from nonamer to pentacamer were synthesized by the phosphotriester method. Aliquots of segments 2 to 6 and 7 to 10 were phosphorylated using T_4 polynucleotide kinase and $\{\gamma\text{-}^{32}P\}$ATP by the procedure of Khorana's group.[40,31] Segments I and II were not phosphorylated to avoid the formation of undesired dimer and other side products in the joining reaction. Four separate bacteriophage T_4 ligase-catalyzed joining reactions were performed. (i) Unphosphorylated segment I (5.9 nmol), ^{32}P-segment 2 (3.0 nmol), and ^{32}P-segment 7 (3.0 nmol) were joined together by base-pair complementation to produce duplex (a), with a yield of 1.8 nmol. (ii) ^{32}P-Segment 3 (3.1 nmol), and ^{32}P-segment 4 (3.3 nmol), and ^{32}P-segment 8 (3.2 nmol) and ^{32}P-segment 9 (3.1 nmol) to give duplex (b), with a yield of 1.3 nmol. (iii) ^{32}P-Segment 5 (3.5 nmol), and ^{32}P-segment 6 (3.5 nmol) were joined in the presence of ^{32}P-segment 10 (3.5 nmol) and unphosphorylated segment 11 (7.0 nmol) to give duplex (c), with a yield of 1.9 nmol. These ligations were carried out at 5°C for 24 hr in 500 $\mu\ell$ of 50 mM Tris-HCl (pH 7.6), 10 mM $MgCl_2$, 10 mM dithiothreitol (DTT), 50 μM ATP using 100 units of bacteriophage T_4 DNA ligase. Before adding ATP and ligase, the reaction mixture was heated at 97°C for 2 min and slowly cooled to 5°C. The three ligation products were then separated by electrophoresis on 15% polyacrylamide. (iv) The isolated duplexes (a, 463 pmol; b, 407 pmol; and c, 406 pmol) were finally joined together to form the DNA fragment VI with a yield of 206 pmol. The reaction was performed as above in a mixture (100 $\mu\ell$) containing 20 units of bacteriophage T_4 DNA ligase, and after 24 hr of incubation at 5°C the ligated fragment was separated by electrophoresis on 15% polyacrylamide. The desired band migrating at 66 bp was isolated and the DNA was phosphorylated enzymatically using bacteriophage T_4 polynucleotide kinase and $\{\gamma\text{-}^{32}P\}$ATP. The phosphorylated product was isolated by gel filtration on a Sephadex® G-50 column.

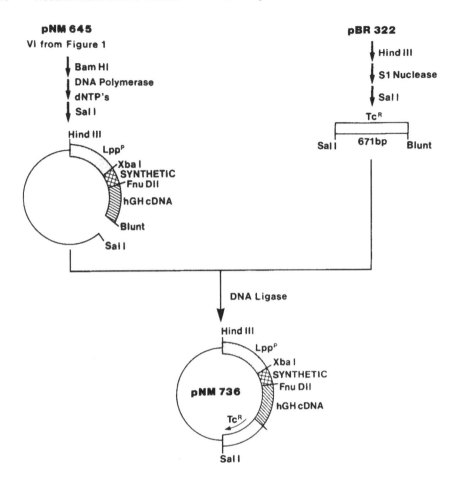

FIGURE 3.   Construction of a plasmid, pNM737, which expresses tetracycline resistance, for the expression of hGH from pNM645. pNM645 was digested with Bam HI and the protruding termini of 2 µg of the resulting DNA were converted to blunt-ended DNA by "filling in" with the use of the Klenow fragment of *E. coli* DNA polymerase I and the four nucleoside triphosphates. The enzyme was then denatured by heating and the mixture was treated with Sal I. The large plasmid fragment was then isolated after its separation on a 1% agarose gel. To isolate the tetracycline gene sequences, 5 µg of pBR322 were digested with Hind III, and the cleaved DNA was then treated with S1 nuclease and then Sal I. The resulting 617-bp fragment was isolated by electrophoresis on a 6% polyacrylamide gel. The plasmid vector (0.05 pmol, 0.2 µg) was ligated with the 617-bp fragment (0.5 pmol, 0.2 µg) using the conditions described above and was then used to transform *E. coli* JA221. Colonies were selected on agar plates containing 100 µg/mℓ of ampicillin and 15 µg/mℓ of tetracycline. Plasmid (pNM736) DNA was isolated from the fragments and found to contain the desired sequence by restriction enzyme analysis. Abbreviations are the same as in Figure 1.

Initial expression of hGH by pNM645 in *E. coli* JA221, detected initially by a modification of the solid phase radioimmunnoassay procedure of Broome and Gilbert,[34-36] was assessed with the use of SDS-polyacrylamide gel analysis of total bacterial cell protein performed according to Laemmli[37] (Figure 4). Analysis of the gels revealed a major protein band of approximately 22,000 daltons. This band was estimated to be about 10% of the total protein, and only a much fainter band of somewhat lower molecular weight was present in preparations of *E. coli* JA221 containing pKEN024. Expression of methionyl hGH by the bacterium harboring pNM736 was found to be as high as that for pNM645 (data not shown).

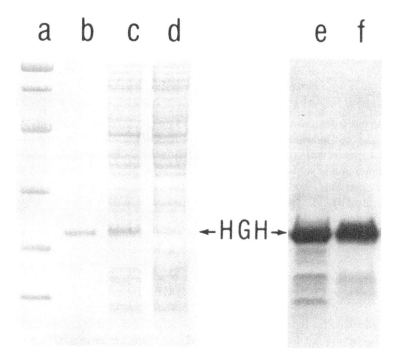

FIGURE 4. Identification of hGH produced in bacteria by SDS-polyacrylamide gel electrophoresis. *a-d* — Proteins stained with Coomassie brilliant blue. Slot *a* contains protein standards of molecular weights 92,500, 66,200, 45,000, 31,000, 21,500, and 14,400 daltons; slot *b* contains purified hGH standard (Cambridge Nuclear); slot *c* contains a cell lysate of JA221/pNM645; and slot *d* contains a cell lysate of JAA221/pKEN024. The crude cell lysates were prepared by growing cells in brain heart infusion broth containing ampicillin at 50 μg/mℓ followed by lysis in 2% SDS and 5% beta-mercaptoethanol. The samples were separated on a 12.5% polyacrylamide slab gel.[37] Slot *e* is an autoradiogram of a whole cell lysate of JA221/pNM645, prepared and separated as in previous samples in which proteins were transferred electrophoretically to a nitrocellulose membrane and incubated with antiserum to hGH (Cambridge Nuclear). After appropriate washes, the membrane was incubated with ^{125}I-protein A and subsequently exposed to film. Slot *f* contains bacterially produced hGH after partial purification as described in the text and run in parallel with the material in slot *e*.

The quantity of hGH in extracts of bacteria harboring pNM645 was measured by radioimmunoassay[38] and found to be at least 2 million molecules/cell. The methionyl hGH was partially purified from 500 gm of *E. coli*. Cells were first extracted with 8 *M* urea and 1% Triton X-100. The debris was removed by centrifugation and the supernatant medium containing the methionyl hGH was fractionated on a Whatman® DE52 column. The peak fractions, as determined by radioimmunoassay, were pooled and subjected to isoelectric precipitation. This material was further purified on a Whatman SE53 column. The peak methionyl hGH-containing fractions were determined by radioimmunoassay and the material was concentrated by isoelectric precipitation or ultrafiltration. That methionyl hGH, rather than hGH, was actually made was verified by comparing the properties of the bacterially synthesized material with that of authentic hGH with the use of high pressure liquid chromatography.

The biological activity of the recovered methionyl hGH was determined by measurement of the proximal epiphyseal cartilage width in hypophysectomized female rats according to the method of Greenspan et al.[39] This material was found to have activity similar to that of hGH obtained from cadavers. Subsequently, this bacterially synthesized material has been purified to homogeneity at the Lilly Research Laboratories and shown to be at least as active on a per mole basis as authentic hGH.

V. CONCLUDING REMARKS

The present work demonstrates that the cloning vector containing the lpp gene machinery can be used to produce large amounts of hGH in *E. coli*. Comparative experiments in our laboratories (unpublished) suggest that the lpp promoter is approximately equal to the *E. coli* trp promoter and is considerably more efficient than the lac promoter for production of human growth hormone. Although our approach and that of Goeddel et al.[14] yield methionyl hGH instead of hGH, the presence of this extra amino acid does not appear to inhibit the biological activity of the hormone. Thus, the data confirm that several different approaches can be used to produce biologically active hGH and that these can be employed to obtain an unlimited supply of this hormone.

ACKNOWLEDGMENTS

We thank Mr. Charles Brush, Mr. Blair Macphail, and Mr. Jim West for their assistance in the synthesis of oligonucleotides, Mr. Richard Van Frank for purifying hGH and Dr. Carl Shaar for testing its biological activity. We also thank Dr. J. Paul Burnett for his interest and helpful suggestions in various aspects of this project.

REFERENCES

1. **Raben, M. S.,** Treatment of a pituitary dwarf with human growth hormone, *J. Clin. Endocrinol.*, 18, 901, 1958.
2. **Goodman, H. G., Grumbach, M. M., and Kaplan, S. L.,** Growth and growth hormone. II. A comparison of isolated growth-hormone deficiency and multiple pituitary-hormone deficiencies in 35 patients with idiopathic hypopituitary dwarfism, *N. Engl. J. Med.*, 278, 57, 1968.
3. **Escamilla, R. F.,** Clinical studies of human growth hormone in children with growth problems, in *Hormonal Proteins and Peptides*, Vol. 3, Li, C. H., Ed., Academic Press, New York, 1975, 147.
4. **Raiti, S., Ed.,** *Proceedings of the NIAMDD Symposium*, Baltimore, Washington, D.C. DHEW, Publication no. NIH 74-612, 1973.
5. **Rudman, D., Kutner, M. H., Rogers, C. M., Lubin, M. F., Fleming, G. A., and Bain, R. P.,** Impaired growth hormone secretion in the adult population: relation to age and adiposity, *Proc. Natl. Acad. Sci. U.S.A.*, 67, 1361, 1981.
6. **Ullrich, A., Shine, J., Chirgwin, J., Pictet, R., Tischer, E., Rutter, W. J., and Goodman, H. M.,** Rat insulin genes: construction of plasmids containing the coding sequences, *Science*, 196, 1313, 1977.
7. **Seeburg, P. H., Shine, J., Martial, J. A., Baxter, J. D., and Goodman, H. M.,** Nucleotide sequence and amplification in bacteria of structural gene for rat growth hormone, *Nature*, 270, 486, 1977.
8. **Shine, J., Seeburg, P. H., Martial, J. A., Baxter, J. D., and Goodman, H. M.,** Construction and analysis of recombinant DNA for human chorionic somatomammotropin, *Nature*, 270, 494, 1977.
9. **Seeburg, P. H., Shine, J., Martial, J. A., Ivarie, R. D., Morris, J. A., Ullrich, A., Baxter, J. O., and Goodman, H. M.,** Synthesis of growth hormone by bacteria, *Nature*, 276, 795, 1978.
10. **Miller, W. L., Martial, J. A., and Baxter, J. D.,** Molecular cloning of DNA complementary to bovine growth hormone mRNA, *J. Biol. Chem.*, 255, 7521, 1980.
11. **Martial, J. A., Hallewell, R. A., Baxter, J. D., and Goodman, H. M.,** Human growth hormone: complementary DNA cloning and expression in bacteria, *Science*, 205, 602, 1979.
12. **Villa-Komaroff, L., Efstratiadis, A., Broome, S., Lomedico, P., Tizard, R., Naber, S. P., Chick, W. L., and Gilbert, W.,** A bacterial clone synthesizing proinsulin, *Proc. Natl. Acad. Sci. U.S.A.*, 75, 3727, 1978.
13. **Fraser, T. H. and Bruce, B. J.,** Chicken ovalbumin is synthesized and secreted by *Escherichia coli*, *Proc. Natl. Acad. Sci. U.S.A.*, 75, 5936, 1978.

14. **Goeddel, D. V., Heyneker, H. L., Hozumi, T., Arentzen, R., Itakura, K., Yansura, D. G., Ross, M. J., Miozzari, G., Crea, R., and Seeburg, P. H.,** Direct expression in *Escherichia coli* of a DNA sequence coding for human growth hormone, *Nature,* 281, 544, 1979.

15. **Seeburg, P. H., Sias, S., Adelman, J., DeBoer, H. A., Hayflick J., Jhurani, P., Goeddel, D. V., and Heyneker, H. L.,** Efficient bacterial expression of bovine and porcine growth hormone, *DNA,* 2, 37, 1983.

16. **DiRienzo, J. M., Nakamura, K., and Inouye, M.,** The outer-membrane proteins of gram-negative bacteria: biosynthesis, assembly, and functions, *Ann. Rev. Biochem.,* 47, 481, 1978.

17. **Halegoua, S. and Inouye, M.,** in *Bacterial Outermembranes: Lipogenesis and Functions,* Inouye, M., Ed., New York, John Wiley & Sons, 1979, 67.

18. **Inouye, M.,** Lipoprotein of the outer membrane of *Escherichia coli,* in *Biomembranes,* Vol. 10, Manson, L. A., Ed., New York, Plenum Press, 1979, 141.

19. **Inouye, S., Wang, S., Sekizawa, J., Halegoua, S., and Inouye, M.,** Amino acid sequence for the peptide extension on the prolipoprotein of the *Escherichia coli* outer membrane., *Proc. Natl. Acad. Sci. U.S.A.,* 74, 1004, 1977.

20. **Nakamura, K., Pirtle, R. M., Pirtle, I. L., Takeishi, K., and Inouye, M.,** Messenger ribonucleic acid of the lipoprotein of the *Escherichia coli* outer membrane. II. The complete nucleotide sequence., *J. Biol. Chem.,* 255, 210, 1980.

21. **Nakamura, K. and Inouye, M.,** DNA sequence of the *Serratia marcescens* lipoproptein gene, *Proc. Natl. Acad. Sci. U.S.A.,* 77, 1369, 1980.

22. **Yamagata, H., Nakamura, K., and Inouye, M.,** Comparison of the lipoprotein gene among the *Enterobacteriaceae* DNA sequence of *Erwina amylovora* lipoprotein gene, *J. Biol. Chem.,* 256, 2194, 1981.

23. **Lee, N., Nakamura, K., and Inouye, M.,** Expression of the *Serratia marcescens* lipoprotein gene in *Escherichia coli, J. Bacteriol.,* 146, 861, 1981.

24. **Nakamura, K. and Inouye, M.,** DNA sequence of the gene for the outer membrane lipoprotein of *E. coli:* an extremely AT-rich promoter, *Cell,* 18, 1109, 1979.

25. **Nakamura, K., Masui, Y., Inouye, M.,** Use of a *lac* promoter-operator fragment as a transcriptional control switch for expression of the constitutive *lpp* gene in *Escherichia coli, J. Molec. Appl. Genet.,* 1, 289, 1982.

26. **Wensink, P. C., Finnegan, D. J., Donelson, J. E., and Hogness, D. S.,** A system for mapping DNA sequences in the chromosomes of *Drosophila melanogaster, Cell,* 3, 315, 1974.

27. **Maxam, A. M. and Gilbert, W.,** A new method for sequencing DNA, *Proc. Natl. Acad. Sci. U.S.A.,* 74, 560, 1977.

28. **Clewell, D. B. and Helinski, D. R.,** Supercoiled circular DNA protein complex in *E. coli:* purification and induced conversion to an open circular DNA form, *Proc. Natl. Acad. Sci. U.S.A.,* 62, 1159, 1969.

29. **Birmboim, H. C., and Doly, J.,** A rapid alkaline extraction procedure for screening recombinant plasmid DNA, *Nucl. Acids Res.,* 7, 1513, 1979.

30. **Brown, E. L., Belagaje, R., Ryan, M. J., and Khorana, H. G.,** Chemical synthesis and cloning of a tyrosine tRNA gene, *Meth. Enzymol.,* 68, 109, 1979.

31. **Sekiya, T., Besmer, P., Takeya, T. and Khorana, H. G.,** Total synthesis of the structural gene for the precursor of a tyrosine suppressor transfer RNA from *Escherichia coli.* 7. Enzymatic joining of the chemically synthesized segments to form a DNA duplex corresponding to the nucleotide sequence 1-26, *J. Biol. Chem.,* 251, 634, 1976.

32. **Narang, S. A., Hsiung, H. M., and Brousseau, R.,** Improved phosphotriester method for synthesis of gene fragments, *Meth. Enzymol.,* 68, 90, 1979.

33. **Crea, R., Kraszewski, A., Hirose, T., and Itakura, K.,** Chemical synthesis of genes for human insulin, *Proc. Natl. Acad. Sci. U.S.A.,* 75, 5765, 1978.

34. **Broome, S. and Gilbert, W.,** Immunological screening method to detect specific translation products, *Proc. Natl. Acad. Sci. U.S.A.,* 75, 2746, 1978.

35. **Hitzeman, R. A., Chinault, A. C., Kingsman, A. J., and Carbon, J.,** Detection of *E. coli* clones containing specific yeast genes by immunological screening, in *Eukaryotic Gene Regulation* (ICN-UCLA Symposia on Molecular and Cellular Biology,) Vol 14, Axel, R., Maniatis, T., Cox C. F., Eds., Academic Press, New York, 1979, 57.

36. **Erlich, H. A., Cohen, S. N., and McDevitt, H. O.,** A sensitive radioimmunoassay for detecting products translated from cloned DNA fragments, *Cell,* 13, 681, 1978.

37. **Laemmli, U. K.,** Cleavage of structural proteins during the assembly of the head of bacteriophage T$_4$, *Nature,* 227, 680, 1970.

38. **Twomey, S. L., Beattie, J. M., and W. u, G. T.,** Shortened radioimmunoassay for human growth hormone, *Clin. Chem.,* 20, 389, 1974.

39. **Greenspan, F. S., Li, C. H., Simpson, M. E., and Evans, H. M.,** Bioassay of hypophyseal growth hormone: the tibia test, *Endocrinology,* 45, 455, 1949.

40. **Nakamura, K. and Inouye, M.,** unpublished results.

Chapter 9

BIOLOGICAL ACTIONS IN HUMANS OF RECOMBINANT DNA SYNTHESIZED HUMAN GROWTH HORMONE

Raymond L. Hintz

TABLE OF CONTENTS

I. INTRODUCTION

Human growth hormone (hGH) is a 191 amino acid polypeptide hormone which is secreted by the pituitary gland and has profound effects on skeletal growth in the immature human. It is clear that hGH also plays a role in the modulation of carbohydrate metabolism. For the past 25 years human growth hormone has been in use for the treatment of certain causes of short stature, especially growth hormone deficiency, and for the treatment of some cases of hypoglycemia.[1] However, unlike other peptide hormones used in clinical medicine such as insulin, the homologous growth hormones from most other species were discovered to be essentially inactive in humans.[2] Only the GH derived from the pituitary glands of monkeys and humans showed significant biological action in the clinical setting. This made the use of pituitary glands from porcine and bovine sources impossible, and dictated the use of human pituitary glands obtained at autopsy as the only practical source available for GH to use therapeutically. This species specificity severely limited the supply of hGH available for clinical use. In the United States the collection of pituitary glands, purification, and distribution of hGH for research for use in patients has been centralized in the National Pituitary Agency. Other countries also have developed national programs for the use of hGH, and at least two pharmaceutical manufacturers have been able to obtain enough pituitary glands to be able to market commercial preparations of hGH. None of these efforts were able to overcome the basic shortage of supply, and for all practical purposes research and treatment has been limited to hypopituitary dwarfism. Even with this very restricted use of hGH, there are still a significant number of children who might benefit from hGH treatment who are not being treated. Furthermore, the dosage of hGH used in treatment of GH deficiency has frequently been determined more by the supply than by any other consideration. There has been very little research done on the potential uses of hGH in other causes of short stature, catabolic states, or disorders of metabolism where pharmacologic doses of hGH may conceivably be of benefit.

Because of the limited supply and potential use for hGH, attempts were made to synthesize short segments of hGH in hopes that they would be biologically active,[3] and even to synthesize hGH totally by solid phase methods.[4] However, no fragment has yet been found that fully duplicates hGH action, and the yield of active hGH after total solid phase synthesis was too low to be a practicable approach. Therefore, hGH was a logical early target for synthesis by recombinant DNA techniques. The cloning and expression of the hGH gene has been discussed in the preceding chapter. The approach used at Genentech, Inc. to produce significant quantities of synthetic hGH was to isolate the mRNA of hGH and make a complimentary DNA copy of it. An endonuclease cleavage was done, and a synthetic leader sequence including the appropriate control signals annealed into place. This partially synthetic hGH gene was then inserted into a plasmid carrier and directly expressed in *Escherichia coli*.[5] The biosynthetic hGH produced by the bacteria has exactly the same amino acid sequence as pituitary hGH, with the addition of a methionyl residue at the N-terminal end. This methionyl residue appears to be necessary for efficient direct expression in the *E. coli* system. Adequate amounts of the recombinant DNA synthesized hGH have been produced so that extensive biological testing has been carried out in both animal models and in humans.

Animal testing in rats and rabbits showed that the synthetic hGH was biologically active in animals, and equipotent to pituitary hGH. Furthermore, the recombinant DNA synthetic hGH was remarkably free of side effect in the acute animals studies.[6] Longer term treatment in animals has led to the development of antibodies and loss of growth stimulation. This was expected from previous studies on the action of hGH in other species, and there was again no difference detected between the synthetic hGH and the pituitary hGH controls. In addition to the animal studies, we were able to show that the synthetic hGH was indistinguishable from pituitary hGH in its ability to bind to and down-regulate the human hGH receptor site in vitro.[7] With this background information it was possible to extend the studies of synthetic hGH to humans. The human studies were planned to occur in three stages:

1.   Short-term tests of acute biological action and safety in adult volunteers.
2.   Long-term tests of efficacy and safety in immature patients with growth hormone deficiency.
3.   Exploration of the therapeutic value of synthetic hGH in other clinical situations.

The short-term tests in humans have been completed,[8,9] and will be discussed in detail in this chapter. The long-term tests of efficacy are underway but not yet complete, and will therefore be described in only a preliminary way. The exploration of other potential uses for synthetic hGH has not yet begun, but the outlines of the possible future therapuetic uses of the synthetic hormone will be discussed.

II. SHORT-TERM HUMAN TESTS OF SYNTHETIC METHIONYL hGH

Synthetic methionyl hGH was produced by the previously described techniques.[5] Human pituitary hGH for control studies was purchased from Serono Laboratories, Braintree, Massachusetts. The two hGH preparations were compared for safety and short-term biological effects in a randomized crossover design. Male volunteers were assigned to treatment groups in a random fashion, and throughout the study they were kept unaware of which hormone was being injected. The physicians involved in the assessment of symptoms were also kept unaware of which hormone the volunteers were receiving. After an initial physical examination, each volunteer had blood drawn for general laboratory screen (SMAC-20), cholesterol, triglycerides, and somatomedin-C (SM-C). In addition, a glucose tolerance test with insulin levels was performed on each volunteer. Following these baseline studies, each volunteer received an intramuscular injection of 0.125 mg/kg (average dose = 8 mg) of either pituitary hGH or the synthetic methionyl hGH. The injections of hGH were given daily for a total of four injections. SM-C was assessed daily, 18 hr after each hGH injection. After one injection, hGH pharmacokinetics were determined in ten patients by hGH radioimmunoassay at 0, 1, 2, 4, 8, and 20 hr. In addition, the volunteers completed a symptom check list and had a physical examination daily. After the initial phase, all subjects had a medication-free period of 10 days. The treatment groups then received the alternative hGH preparation according to the same schedule as before.

Twenty-two volunteers were studied. In the initial group of twelve volunteers it became apparent, when the code was broken, that there was a statistically significant increase of local pain at the injection site, generalized malaise, an increase in segmented neutrophils and bands, and a drop in serum iron during the synthetic methionyl hGH treatment. Multiple in vivo animal tests, both before and after this first human test, failed to reveal any similar toxic effect of treatment with the synthetic met-hGH. In addition, the standard tests for endotoxin and pyrogens were consistently negative. However, an endogenous pyrogen contamination of the synthetic preparation was revealed by the assay of Dinarello et al.[10] This endogenous pyrogen disappeared after an additional chromatographic step was introduced in the purification procedure for the synthetic hGH. Subsequently, the study was repeated on 10 additional volunteers with the new preparation of synthetic methionyl hGH. The same protocol for study was used as in the first acute human study. In the repeat study there was no statistically significant difference found between the response to pituitary hGH and synthetic methionyl hGH in terms of symptoms, neutrophilia, or serum iron. There were no differences between the two studies in the hGH related responses measured, and the data from all twenty-two patients were pooled for analysis.

Figure 1 shows the pharmacokinetics of 0.125 mg/kg of pituitary hGH and synthetic hGH after intramuscular injection. With both hGH preparations plasma hGH levels rose to supra-

GH PHARMACOKINETICS

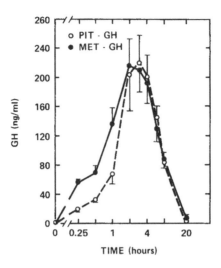

FIGURE 1.   Growth hormone pharmacokinetics.

physiological levels and peaked at 3 to 4 hours. The hGH levels remained elevated above the basal level even in the 20 hour sample. There was a small but statistically significant difference between the hGH levels achieved in the two treatment groups at the early time points, with the levels achieved after synthetic hGH being consistently higher. However, there was no difference between the groups in peak levels, area under the hGH curve, or apparent dissappearance rate.

It is clear that many of the in vivo effects of growth hormone on cartilage and bone are not direct effects of hGh itself, but are instead mediated by a group of plasma peptides under hGH control known collectively as the somatomedins.[11] Two of the human somatomedins have been sequenced and shown to have strong structural homologies to proinsulin.[12] They have been renamed insulin-like growth factor (IGF) I and II. Although the terminology is still confusing, it appears that IGF-I and SM-C are identical molecules, and that the level of SM-C in the plasma is closely linked to hGH action. We, therefore, chose to monitor SM-C levels as the best single reflector of hGH biological action in vivo. The administration of these supra-physiological amounts of hGH caused a prompt and steady rise in plasma SM-C in the volunteers (Figure 2). By the fifth day of the study, the levels of SM-C were clearly in the supra-physiologic range, and comparable to what is seen in acromegaly. There was no significant difference between pituitary hGH and synthetic methionyl hGH in their ability to increase the SM-C levels.

One of the most consistent effects of growth hormone in vivo has been the increase in protein synthesis with concomitant drops in nitrogen excretion and blood urea nitrogen. In both the pituitary hGH and synthetic hGH treatment groups there was a striking drop in blood urea nitrogen ($P < 0.001$) after the four injections, and once again there was no significant difference between the two treatment groups (Figure 3). This indicates that the short-term effect on nitrogen retention is just as potent in the synthetic methionyl hGH as it is in pituitary hGH.

Another classic effect of growth hormone has been the lipolytic effect, both in vitro and in vivo. Both pituitary hGH and synthetic hGH caused an increase in circulating serum triglycerides (Figure 4). This difference is significant when the response of each patient is compared with his own baseline ($p < 0.05$). In addition, both preparations caused a small

△ SM-C (RIA) WITH GH Rx

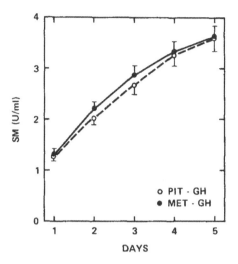

FIGURE 2. Changes in SM-C after growth hormone administration.

△ BUN WITH GH Rx

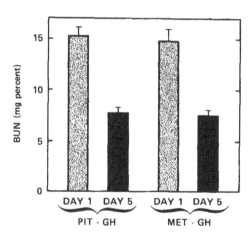

FIGURE 3. Blood urea nitrogen before and after 4 days of growth hormone administration.

but significant (p<0.05) drop in serum cholesterol (Figure 5). There were again equal effects of the synthetic methionyl hGH and the pituitary hGH.

Growth hormone has long been known to be involved in the modulation of sensitivity to insulin. Increases in insulin resistance have been associated with increases in growth hormone both in hypopituitarism treated with hGH,[13] and in acromegaly.[14] In our volunteers the glucose tolerance tests showed a clear deterioration in blood glucose response after 4 days of treatment (Figure 6). The fasting plasma glucose levels were modestly, but significantly, elevated after pituitary hGH (105.9 ± 3.0 vs. 96.6 ± 2.9 mg/dℓ) and methionyl hGH (105 ± 3.3 vs. 96.2 ± 1.5 mg/dℓ). The elevation 60 minutes after the administration of 100 grams of glucose was much more apparent, with a more than 150% increase in glucose

△TRIGLYCERIDES WITH GH Rx

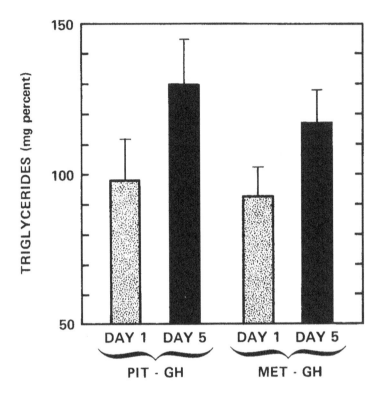

FIGURE 4.   Triglycerides before and after 4 days of growth hormone administration.

levels after growth hormone treatment. Even more striking were the changes seen in the insulin response to the glucose tolerance tests after 4 days of hGH therapy. The basal insulin levels were elevated more than threefold after the four days of hGH injection, and the area under the insulin curve essentially doubled (Figure 7). These changes are good evidence that marked insulin resistance is developing with the supra-physiologic doses of hGH, very similar to the changes seen in acromegaly.[14] This is not due to changes in the insulin receptor, since there were no significant changes in the insulin receptor sites during these studies.[9] At no time were any differences seen in glucose, insulin, or insulin receptor sites between the pituitary hGH and the synthetic hGH groups.

Thus, from these short-term studies it appears that the synthetic hGH has full biological activity judged by its actions on SM-C generation, protein anabolism, and changes in fat and carbohydrate metabolism. Although no formal dose response comparison was done in these acute human studies, it appears from the comparison of equal dosages that the synthetic and pituitary hormones are equipotent. These studies also provide a good example of the need for caution in testing products of recombinant DNA technology. Chemically minuscule amount of bacterial contamination that are not apparent on either animal testing or conventional gel electrophoresis may still be of importance clinically. Testing for endogenous pyrogen activity may be of great help in testing products before clinical studies are undertaken, but there appears to be no real substitute for carefully done acute human safety studies before longer term human trials are undertaken.

△ CHOLESTEROL WITH GH Rx

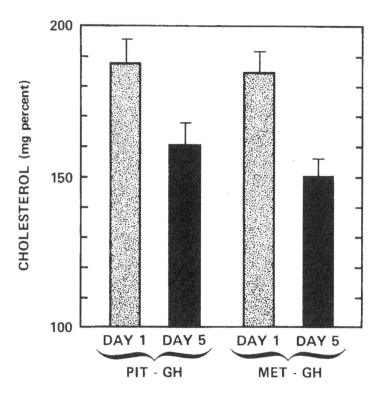

FIGURE 5. Cholesterol before and after 4 days of growth hormone administration.

III. LONG-TERM HUMAN STUDIES

Longer term studies are now underway to determine if the synthetic hGH is efficacious in the treatment of growth hormone deficiency. This is being conducted as a multicenter test in children with well-documented hGH deficiency. At the time of this writing, these studies are still in progress, and therefore they cannot be discussed in detail. However, it seems clear that all of the hypopituitary patients are showing an increased growth rate to treatment with synthetic hGH, and no major toxic side effects have occurred. Assuming that these studies in hypopituitary patients are successfully completed, the synthetic methionyl hGH will be released for the treatment of short stature due to hypopituitarism, and for further exploration of other potential therapeutic uses of growth hormone treatment. Preliminary data has suggested that hGH might have a role in the treatment of such disorders as cachexia, burns, osteoporosis, and delayed healing of fractures.[15] The supply of hGH has so far been so limited that it has not been possible to organize appropriate clinical trials with adequate numbers of patients. With the potentially unlimited availability of synthetic hGH, such studies can now be done. Another area in which it has been proposed that hGH may be of some potential therapuetic use is in the treatment of short stature in children that is not due to growth hormone deficiency. The data of one investigator suggest that a significant proportion of these otherwise normal children may respond to hGH with an increase in their growth rates.[16] Only with the availability of synthetic hGH can carefully controlled long-term studies of these patients be done. The crucial questions that must be answered are whether there is any way to predict which children might respond to hGH therapy, and whether any short-term response in growth rate can be sustained so that there is an actual increase in the final height of these patients.

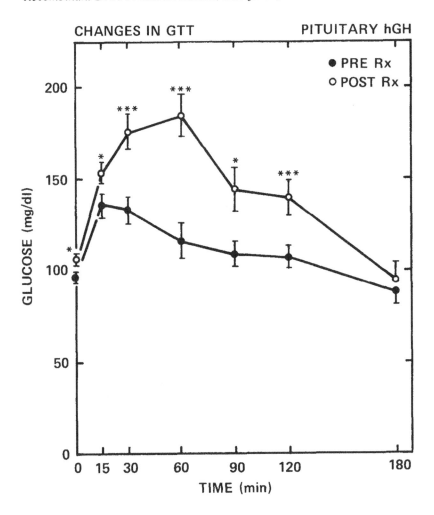

FIGURE 6.   Plasma glucose levels during GTT before and after 4 days of growth hormone administration.

It is easy to create the picture of an "Orwellian" world in which each child's height is tailored to the an arbitrary standard or the parents' wishes. However, human growth hormone therapy now involves a considerable amount of inconvenience, expense, and long-term commitment from the parents and patients, and it seems likely to remain so even after the availability of hGH is increased by the synthetic preparation. Thus, there are some built in natural safeguards to prevent widespread abuse of hGH. Nevertheless, it is crucial that good clinical studies be carried out on the potential expanded clinical uses of hGH so that there are reasonable guidelines available for its use.

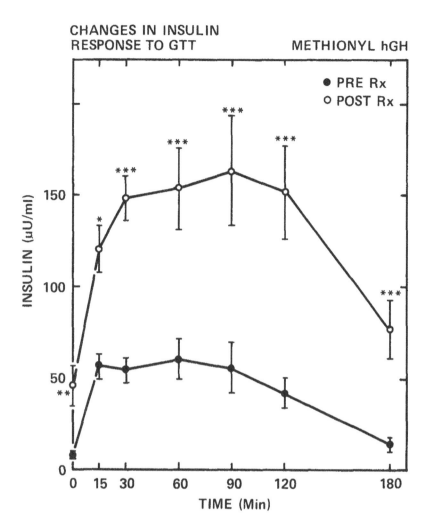

CHANGES IN INSULIN
RESPONSE TO GTT   METHIONYL hGH

FIGURE 7.   Plasma insulin levels during GTT before and after 4 days of growth hormone administration.

REFERENCES

1. **Raben, M. S.,** Treatment of a pituitary dwarf with human growth hormone, *J. Clin. Endocrinol. Metab.,* 18, 901, 1958.
2. **Bergenstal, D. M. and Lipset, M. B.,** Metabolic effects of human growth hormone and growth hormone of other species in man, *J. Clin. Endocrinol. Metab.,* 29, 1427, 1960.
3. **Nial, H. D. and Tregear, G. W.,** Structural and synthetic studies of hGH fragments. in *Advanceces in Human Growth Hormone Research.,* S. Raiti, Ed., DHEW 74-612, 394, 1974.
4. **Lee, C. H. and Yamashiro, D.,** Synthesis of a growth hormone related protein, *J. Amer. Chem. Soc.,* 92, 7608, 1970.
5. **Goeddel, D. V., Heyrecer, H. L., Hozumi, T., Arenteen, R., Itakura, K., Vansura, D. G., Ross, M. J., Miozagri, G., Crea, R., and Seeburg, P. H.,** Direct expression in *E. coli* of a DNA sequence coding for human growth hormone, *Nature,* 281, 544, 1979.

6. **Olson, K. C., Fenno, J., Lin N., Harkins, R. N., Snider, C., Kohr, W. H., Ross, M. J., Fodge, D., Drandor, G., and Stebbing, N.,** Purified human growth hormone from *E. coli* is biologically active, *Nature,* 293, 408, 1981.

7. **Rosenfeld, R. G., Aggarwal, B. B., Hintz, R. L., and Dollar, L. A.,** Recombinant DNA-derived methionyl human growth hormone is similar in membrane binding properties to human pituitary growth hormone, *Biochem. Biophys. Res. Commun.,* 106, 202, 1982.

8. **Hintz, R. L., Rosenfeld, R. G., Wilson, D. M., Bennett, A., Fenno, J., McClellan, B., and Swift, R.,** Biosynthetic methionyl human growth hormone is biologically active in adult man, *Lancet,* 1, 1276, 1982.

9. **Rosenfeld, R. G., Wilson, D. M., Dollar, L. A., Bermett, A., and Hintz, R. L.,** Both human pituitary growth hormone and recombinant DNA-derived human growth hormone cause insulin resistance at a postreceptor site, *J. Clin. Endocrinol. Metab.,* 54, 1033, 1982.

10. **Dinarello, C. A. and Wolff, S. M.,** Pathogenesis of fever in man, *N. Engl. J. Med.,* 298, 607, 1978.

11. **Van Wyk, J. J., Underwood, L. E., Hintz, R. L., Clemmons, D. R., Voina, S. J., and Weaver, R. P.,** The somatomedins: a family of insulin-like hormones under growth hormone control, *Rec. Prog. Hormone Res.,* 30, 259, 1974.

12. **Zapf, J., Rinderknecht, E., Humbel, R. E., and Froesch, E. R.,** Nonsupressible insulin-like activity from human serum: recent accomplishments and their physiological implications, *Metabolism,* 27, 1803, 1978.

13. **Underwood, L. E., Van den Brande, J. L., Antony, G. J., Voina, S. J., and Van Wyk, J. J.,** Islet cell function and glucose homeostasis in hypopituitary dwarfism: synergism between growth hormone and cortisone, *J. Pediatr.,* 82, 28, 1973.

14. **Beck, P., Schlach, D. S., Parker, M. L., Kipnis, D. M., and Daughaday, W. H.,** Correlative studies of growth hormone and insulin plasma concentrations with metabolic abnormalities in acromegaly, *J. Lab. Clin. Med.,* 66, 366, 1965.

15. **Rudman, D.,** Potential Indications for human growth hormone, in *Insulins, Growth Hormone, and Recombinant DNA Technology.* J. L. Gueriguian, Ed. Raven Press, N.Y., 1981, 161.

16. **Rudman, D., Kutner, M. H., Goldsmith, M. A., Kenney, J., Jennings, H., and Bain, R. P.,** Further observations on four subgroups of normal variant short stature, *J. Clin. Endocrinol. Metab.,* 51, 1378, 1980.

Chapter 10

THE NIH GUIDELINES FOR RESEARCH INVOLVING RECOMBINANT DNA MOLECULES

Elizabeth Ann Milewski

TABLE OF CONTENTS

I. INTRODUCTION

The technique known as recombinant DNA or gene cloning was first developed in the early 1970s. Using this technique, pieces of DNA from virtually any source — no matter how unrelated the sources — may be recombined and subsequently introduced into living cells. The recipient cells would potentially be capable of maintaining and expressing this information, and passing the information on to their progeny.

The development of this incredibly powerful technique led thoughtful scientists, aware of the history of use and abuse of similar breakthroughs in chemistry and physics, to raise questions about possible hazards. These discussions began before the technique had been used enough to permit the possibilities, complications, and potential hazards of application of the technique to be evaluated from a strong data base. The initial discussions, however, raised enough concern that scientists asked, first, for a moratorium on some types of experiments, and secondly, for the development of guidelines for the use of recombinant DNA technology.

In response, the National Institutes of Health (NIH) developed and implemented Guidelines for Research Involving Recombinant DNA Molecules. Over the years, the Guidelines have been modified to reflect changing perceptions on the application of recombinant DNA technology. The changing perceptions can be seen most readily in the specifications and uses of various host-vector systems, e.g., the EK1 and EK2 systems and viral vectors. The history of viral vectors and changing perceptions concerning their use have been covered in depth in the Appendix (Chapter 10A).

During the period of their existence, the Guidelines have affected the development and application of the technology. Some examples of products researched and developed under the Guidelines, human insulin, growth hormone, and interferon, are the subjects of this book. The evolution of the Guidelines is discussed in this article, with some attempt made at describing the features of the Guidelines specifically affecting the industrialization of biotechnology. The article also looks briefly at the commercial development of human insulin, human interferon, and human growth hormone under the Guidelines.

II. THE FIRST ISSUANCE OF THE GUIDELINES IN THE UNITED STATES

In 1973, the promise of the recombinant DNA technique and questions of conjectural risk were discussed at a Gordon Conference session. The participants in that discussion voted that a letter expressing their concerns be sent to the National Academy of Sciences (NAS). The letter, which appeared in *Science* in July 1973,[1] suggested that the National Academy of Sciences "consider this problem and recommend specific actions or guidelines."

In response, the National Academy appointed a distinguished expert panel to evaluate the issue. After a year of discussion, the committee issued the following recommendations:[20]

- "First, and most important, that until the potential hazards of such recombinant DNA molecules have been better evaluated or until adequate methods are developed for preventing their spread, scientists throughout the world join with the members of this committee in voluntary deferring (certain) experiments."
- Second, the letter called for the National Institutes of Health to establish an Advisory Committee for "devising guidelines to be followed by investigators working with potentially hazardous recombinant molecules."
- Third, the committee called for an international conference of experts to discuss the issues.

An international conference was convened in February 1975 at the Asilomar Conference Center in California. The final report of the conference suggested that recombinant DNA experiments may proceed, but biological and physical containment standards should be assigned to each experimental protocol on the basis of perceived hypothetical risk. During this time, the NIH had formed the Recombinant DNA Advisory Committee (RAC) in response to the NAS expert panel recommendation. RAC first met on the day following the Asilomar meeting and proceeded to develop guidelines for recombinant DNA research. These guidelines suggested biological and physical containment standards based on perceived hypothetical risk. The guidelines were presented to Dr. Donald S. Fredrickson, then Director of the NIH, who called a meeting of his Director's Advisory Committee to discuss the proposal. Many distinguished scientific and public representatives were invited to participate and comment and still other witnesses came voluntarily to testify. Questions, observations, and comments on the proposed guidelines were then presented to RAC for further evaluation. After an analysis of the proposed guidelines, and of public comments and observations, Dr. Fredrickson issued the initial version of the NIH Guidelines for Research Involving Recombinant DNA Molecules in July 1976.[2]

These Guidelines were binding on investigators whose research was supported by the NIH. Investigators receiving funds from other sources including other federal agencies were not obliged at this point to adhere to the NIH Guidelines. In July 1976, Senators Jacob Javits and Edward Kennedy wrote to President Gerald Ford urging that "every possible measure be explored for assuring that the NIH Guidelines are adhered to in all sectors of the research community." In reply, President Ford described the creation of the Federal Interagency Committee on Recombinant DNA Research. The Committee is composed of all the Federal agencies which might either fund or regulate recombinant DNA research (Table 1) and provides for the communication necessary to coordinate all activities related to recombinant DNA research. The Committee also is responsible for facilitating compliance with a uniform set of guidelines for recombinant DNA research in the public and private sectors and, where warranted, for suggesting administrative or legislative proposals. At the second meeting of the Committee, all the Federal agencies endorsed the NIH Guidelines, and Departments which support or conduct recombinant DNA research agreed to abide by the NIH Guidelines.

III. THE GUIDELINES FROM 1978 TO 1982

Approximately $2^1/_2$ years elapsed between the issuance of the original Guidelines in 1976 and the first major revision in December 1978. During this period a series of scientific and public meetings demonstrated a consensus that the guidelines were needlessly restrictive. To reflect this consensus, the NIH Director's Advisory Committee was called upon in December 1977 to review proposed revisions. Containment conditions for experiments covered by the Guidelines were evaluated at meetings of scientific experts, and a new version of the Guidelines was published for public comment in July 1978.[3] A public hearing chaired by the General Counsel of the Department of Health, Education, and Welfare (then HEW, now HHS) was held in September 1978 to examine this revised document. The Guidelines, modified to reflect these $2^1/_2$ years of discussion, were reissued on December 22, 1978.[4]

As recombinant DNA technology evolves and new information is acquired, the guidelines for research involving recombinant molecules should be responsive to changing information and perceptions. A mechanism by which to amend and modify their content was incorporated into the Guidelines in the first major revision: A proposal to modify any section or to introduce new provisions into the Guidelines must be published for public comment in the *Federal Register* at least 30 days prior to a RAC meeting. The proposal and any public comments are then discussed by the RAC in open public session. A final decision on the proposal is rendered by the Director, NIH, after consideration of RAC recommendations. Using these procedures, modifications in the Guidelines have been made following every

Table 1

DEPARTMENTS AND AGENCIES COMPOSING THE FEDERAL INTERAGENCY COMMITTEE ON RESEARCH INVOLVING RECOMBINANT DNA ACTIVITIES[a]

Department of Agriculture
Department of Commerce
Department of Defense
Department of Energy
Environmental Protection Agency
Executive Office of the President
Department of Health and Human Services
   Office of the Assistant Secretary for Health
   Centers for Disease Control
   Food and Drug Administration
   National Institutes of Health
Department of the Interior
Department of Justice
Department of Labor
National Aeronautics and Space Administration
National Science Foundation
Nuclear Regulatory Commission
Department of State
Department of Transportation
Arms Control and Disarmament Agency
Veterans Administration

[a] As of December 1980

RAC meeting held since December 1978. Several complete republications have been issued since 1978. All of the republications of the Guidelines are summarized in Table 2. Those republications which occurred between the 1978 reissuance and the current Guidelines are; in January 1980 a revision which substantially modified the status of the host-vector systems used in the majority of recombinant DNA experiments necessitated the reissuance of the Guidelines.[5] In November of 1980, the Guidelines were again reissued when the procedures by which principal investigators register recombinant DNA experiments were modified.[6] In July of 1981 the Guidelines were reissued to reflect a revision exempting from the Guidelines those host-vector sytems used in approximately 80 to 90% of all recombinant DNA experiments.[7]

IV. THE GUIDELINES IN 1982: EVENTS LEADING TO THE CURRENT GUIDELINES

In April 1981, Drs. Allan Campbell and David Baltimore, RAC members, introduced to RAC a proposal[8] to convert the NIH Guidelines into a "code of standard practice." The proposal would have reduced containment to the P1 level for most recombinant DNA experiments with the exception of prohibited experiments. The authors coupled the reduction in containment to the elimination of the punitive aspects of the Guidelines.

Although the RAC did not endorse the Baltimore-Campbell proposal in April 1981, there appeared to be some agreement that a major overhaul of the Guidelines was overdue. It was felt that at the minimum a reevaluation and reorganization of the Guidelines should be undertaken. To this end, RAC urged that a working group be formed to evaluate the status of the NIH Guidelines for Research Involving Recombinant DNA Molecules. The review

Table 2
COMPARISION OF SUBSEQUENT REVISIONS OF THE GUIDELINES TO THE 1976 GUIDELINES[a]

1976 Guidelines (June 1976)
I.   Prohibited experiments specified.
II.   1. Described in great detail four sets of special practices, equipment and laboratory installations (P1, P2, P3, P4).
      2. Defined the concept of biological containment for *E. coli* K-12 systems (EK1, EK2, EK3).
III.   Specification of physical and biological containment levels for different classes of permitted *E. coli* K-12 experiments.
IV.   Specifications of roles and responsibilities of the scientists, the institution, its Institutional Biosafety Committee (IBC), and the NIH.

1978 Guidelines (December 1978)
I.   Certain classes of experiments deemed of the lowest potential hazard exempted from Guidelines.
II.   Biological containment specifications generalized to include other prokaryotic and eukaryotic host-vector systems.
III.   In general experiments were assigned lower containment levels. Experiments with a larger variety of host-vector systems permitted.
IV.   1. Increased representation mandated for local IBCs and RAC.
      2. Modification procedure becomes part of the Guidelines.
      3. Penalties specified for non-compliance.

1980 Guidelines (January 1980 and November 1980)
III.   1. Many experiments assigned lower containment.
      2. Relaxed containment specifications for most frequently utilized host-vector systems *(E. coli* K-12 EK1, *Saccharomyces cerevisiae)*.
IV.   NIH no longer responsible for registration and review of experiments for which containment is clearly specified in the Guidelines (only IBC review required).
VI.   Voluntary compliance provision.

1981 Guidelines (July 1981)
I.   Exempt class of experiments expanded to include the most frequently utilized host-vector systems *(E. coli* K-12 EK1, *S. cerevisiae*, asporogenic *Bacillus subtilis)*.

1982 Guidelines (April 1982 and August 1982)
I.   The term prohibition is no longer used. However three of the prohibitions (drug resistance traits, toxin genes, and deliberate release to the environment) appear in a new Section III-A and require RAC review and approval. Two of the prohibitions (formation of recombinant DNAs from pathogenic organisms and large-scale procedures) are listed in Section III-B and may proceed following IBC review and approval.
II.   1. Reference to HV3 host vector systems deleted.
      2. Section move to Appendices G, H, and I.
III.   1. Greatly simplified. As noted above, new Section III-A contains three of the prohibitions. New Section III-D contains the exemptions. New Section III-B specifies experiments requiring IBC approval before initiation. They are assigned containment levels P1 to P4. New Section III-C includes all experiments not included in Sections III-A, III-B or III-D. These experiments require IBC notification simultaneously with initiation of experiment and can be carried out at P1 containment. Section III-D are those experiments exempt from the Guidelines. Physical containment requirements for some classes of experiments are lowered.
      2. The *CDC Classification of Etiologic Agents on the Basis of Hazard*, the original source of Appendix B, is amended to suit the purposes of the Guidelines.
IV.   1. The responsibilities of the IBC need not be restricted to recombinant DNA.
      2. Requirement that the 20% of the IBC membership not be affiliated with the IBC deleted. The institution would still be required to appoint two non-affiliated members to the IBC.

[a]   Categorized under major section of the Guidelines.

would include but not be limited to: "(a) the present need for the Guidelines in their existing form and procedures, as opposed to a voluntary standard of practice; (b) continued applicability to recombinant DNA technology; (c) currently recommended levels of containment; (d) current processes and procedures impeding or facilitating research and/or industrial application."[21]

A Working Group on Revision of the Guidelines was appointed and convened in June and July 1981. The working group developed a proposal to modify the Guidelines.[9] The proposal recommended a general lowering of containment requirements, and suggested that the mandatory aspects of the Guidelines be maintained. As part of its final report the working group generated the document "Evaluation of the Risks Associated with Recombinant DNA Research",[10] which evaluated scientific information, risk assessment data, and scientific perceptions.

The working group proposal was presented to the RAC in September 1981 and provided a basis for initiating RAC discussion of the status of the Guidelines. During the discussion, Dr. David Baltimore offered a proposal which amalgamated portions of the Baltimore-Campbell proposal[8] with the working group proposal.[9] By modifying Dr. Baltimore's proposal, RAC developed its own proposal for publication in the *Federal Register;* this proposal was published to elicit comment on the concept of converting the Guidelines to a code of standard practice. The proposal appeared in the *Federal Register* in December 4, 1981.[11]

In order to offer RAC and the NIH an alternative proposal, Dr. Susan Gottesman, the chairperson of the Working Group on Revision of the Guidelines, developed a proposal[12] based on the document "Evaluation of the Risks Associated with Recombinant DNA Research."[10] The proposal, offered by Dr. Gottesman, appeared in the *Federal Register* for December 7, 1981.[12]

The two proposals were presented to the RAC for discussion and evaluation in February 1982. Although the proposals differed in several features, the major differences between the proposals were (1) whether the Guidelines would continue to be mandatory for institutions receiving NIH funding, and (2) whether the Institutional Biosafety Committees (IBCs)* would continue their oversight function. At the February 1982 meeting the RAC endorsed the proposal[12] to simplify the Guidelines offered by Dr. Gottesman rather than the proposal[11] to convert the Guidelines to a code of standard practice. The NIH accepted the RAC recommendation to maintain the mandatory nature of the Guidelines and issued revised Guidelines in April 1982.[13]

In August 1982, the Guidelines were reissued to fill the recommendation RAC made in February 1982; that a working group be formed to simplify further and modify the Guidelines.[13] A working group, the Working Group on Revision of the Guidelines, was duly convened and generated a document which was presented to RAC in June 1982. The document refined the April 21, 1982, version of the Guidelines, while changing few of the essentials of the April 21, 1982 Guidelines. The RAC endorsed most of the working group suggestions at the June 25, 1982 meeting. The NIH issued Guidelines based on those recommendations on August 27, 1982.[14] Table 2 compares the salient differences between the six major revisions of the Guidelines as well as the differences between the April 21, 1982 Guidelines and the Guidelines published on August 27, 1982.

V. VOLUNTARY COMPLIANCE BY INSTITUTIONS NOT SUPPORTED BY FEDERAL FUNDS

The original 1976 Guidelines[2] were directed primarily toward investigators located in a university, utilizing federal research monies. Thus, the 1976 Guidelines were silent con-

---

* The NIH Guidelines for Research Involving Recombinant DNA Molecules require that each Institution conducting or sponsoring recombinant DNA research covered by the Guidelines ensure that the research is carried out in full conformity with the provisions of the Guidelines. One specification these institutions must meet is to establish an Institutional Biosafety Committee (IBC). The IBC shall be constituted to collectively have the experience and expertise necessary to assess the safety of recombinant DNA experiments and to assess any potential risk to public health or the environment. At least two members of the committee represent the surrounding community. The IBC is responsible for reviewing recombinant DNA projects for compliance with the NIH Guidelines.

cerning compliance with the Guidelines by industry or other private institutions. Efforts were made in the 95th Congress to legislate national regulation of recombinant DNA. Such legislation would have uniformly affected recombinant DNA research in the public and private sector in the United States. These efforts at national legislation, however, failed. In the absence of national legislation, the NIH, with the endorsement of the Federal Interagency Committee, provided a mechanism for voluntary compliance by the private sector. Part VI of the Guidelines "Voluntary Compliance", which was added in January 1980, formally encourages voluntary compliance by the private sector and specifies how NIH will protect proprietary information voluntarily submitted. No penalties are specified in the Guidelines for those institutions complying voluntarily with the Guidelines.

At this time, 58 companies have registered IBCs with NIH under Part VI of the Guidelines. Included among these 58 are those companies in the forefront in commercially producing human insulin, interferon, and growth hormone by using recombinant DNA techniques. These commercial productions entail the large-scale culturing of recombinant organisms, so how the Guidelines have evolved regarding large-scale procedures will be examined in greater detail in the following section.

VI. LARGE-SCALE APPLICATIONS OF RECOMBINANT DNA TECHNOLOGY UNDER THE GUIDELINES

When the Guidelines were originally formulated in 1975 to 1976, certain categories of recombinant DNA experiments were not to be performed. Among the experiments not to be performed were "large-scale" experiments. When the Guidelines were revised in 1978, a specification was incorporated which prohibited experiments involving more than 10 ℓ of culture in any one volume. The language of the specification, numbered I-D-6, stated that "the following experiments are not to be initiated at this time" and proceeded to prohibit "large-scale experiments . . . with organisms containing recombinant DNAs, unless the recombinant DNAs are rigorously characterized and the absence of harmful sequences established." Exceptions to the prohibition could be obtained if the experiments were expressly approved by the NIH.

Over the years a number of requests for exception to this prohibition were received by NIH, most of these requests were submitted by institutions voluntarily complying with the NIH Guidelines.

VII. PROCEDURES FOR REVIEW OF LARGE-SCALE SUBMISSIONS (1979 TO 1981)

Under procedures implemented in September 1979, RAC began reviewing submissions involving recombinant DNA organisms in large volumes of culture. In these earliest reviews of large-scale experiments, RAC evaluated both the biological properties of the organisms and the physical containment facilities and procedures (see Table 3). In September 1980 after much discussion and debate, RAC determined that it would continue to evaluate the biology of recombinant systems, but that it no longer would review detailed information on large-scale fermentation facilities. Rather, RAC delegated this later responsibility to the local IBCs.

In September of 1981, RAC considered a proposal to modify the language of prohibition I-D-6. This proposal would have changed the review process to be followed when *certain* host-vector systems were used in large-scale procedures; large-scale procedures utilizing the well characterized *E. coli* K-12 EK1, *Saccharomyces cerevisiae*, and asporogenic *Bacillus subtilis* host-vector systems would be reviewed by the local IBC rather than by RAC. The IBCs would evaluate the biology of experiments utilizing these three host-vector systems,

Table 3

PROCEDURES FOR REVIEW OF LARGE SCALE EXPERIMENTS

The following procedures for review of large-scale experiments were adopted by the RAC.

A. For each research project proposing to exceed the 10-liter limit, the applicant shall file a request with the NIH Office of Recombinant DNA Activities (ORDA). The request should include the following information:

1a. The registration document submitted to the local Institutional Biosafety Committee. This should include, or have appended to it, a summary paragraph which describes the proposed project in language that is comprehensible to non-specialists.

2. At statement of the rationate for wishing to exceed the 10-liter limit.

3. A specification of the total volume of the fermentor to be used.

4. Evidence that the recombinant DNAs to be employed in the research have been rigorously characterized and are free of harmful sequences.

5. A description of the applicant's laboratory practices, containment equipment, and facilities relevant to the containment of large volumes of culture.

6. Evidence of the applicant's laboratory practices, containment equipment, and facilities relevant to the containment of large volumes of culture.

7. A description of procedures to be employed for the inactivation and disposal of large volumes of culture.

8. A description of procedures for containing and inactivating accidental spills, should they occur.

B. Each request submitted to ORDA shall be referred to a working group of the NIH Recombinant DNA Advisory Committee for review.

C. Following review and approval by the working group, each request shall be submitted to the entire Recombinant DNA Advisory Committee for review.

D. Following review and approval by the RAC, each request shall be submitted to the Director, NIH, for final review.

E. Applications for large-scale experiments which are submitted by institutions not receiving NIH funds for recombinant DNA research shall be kept confidential (provided the institutions so desire) in accordance with the provisions of the NIH Guidelines for Research Involving Recombinant DNA Molecules and to the extent permitted by law.

F. These procedures may be refined or revised on the basis of discussion and action by the NIH Recombinant DNA Advisory Committee.

as well as the physical facilities to be employed. Such experiments would continue to be covered by the Guidelines, and the IBC would review protocols before initiation of the procedure. Large-scale procedures utilizing other host-vector systems (e.g., *Streptomyces* or sporogenic *Bacillus subtilis*) would continue to be subject to RAC review and NIH approval. RAC recommended acceptance of this proposal to modify prohibition I-D-6, and in October 1981,[15] the NIH officially modified its policy concerning *E. coli* K-12, *S. cerevisiae*, and asporogenic *B. subtilis* host-vector systems used in large-scale procedures. In making this recommendation RAC accepted the argument that potential environmental problems associated with the use of well-characterized organisms in large-scale recombinant DNA production processes appear to be similar to those associated with non-recombinant DNA large-scale fermentations which industry has been performing for many years with an excellent safety record.

The procedures for overseeing large-scale operations with recombinant DNA containing organisms were modified, in a major way, in the April 1982 version[13] of the Guidelines: the authority to determine appropriate containment for most-large scale procedures was delegated to the local IBC. At that time, specification I-D-6, no longer identified as a prohibition, was renumbered III-B-5.

The pertinent section of the April 1982 Guidelines reads as follows:

"III-B-5. *Experiments Involving More than 10 Liters of Culture.* The appropriate containment will be decided by the IBC. Where appropriate, the large-scale containment recommendations of the NIH should be used (45 FR 24968)."*

Investigators must submit to their IBC, prior to initiation of the experiments, " a registration document that contains a description of: (a) the source(s) of DNA, (b) the nature of the inserted DNA sequences, (c) the hosts and vectors to be used, (d) whether a deliberate attempt will be made to obtain expression of a foreign gene, and, if so, what protein will be produced, and (e) the containment conditions specified in these Guidelines. This registration document must be dated and signed by the investigator and filed only with the local IBC. The IBC shall review all such proposals prior to initiation of the experiments."[22]

Table 4 summarizes the changes which have occurred in NIH procedures for evaluating proposals involving the large-scale culturing of organisms containing recombinant DNA.

VIII. CURRENT LARGE-SCALE PROCEDURES UNDER THE GUIDELINES

As of August 27, 1982, procedures involving manipulation and growth of recombinant DNA organisms in greater than 10 ℓ volumes are covered by the NIH Guidelines. Responsibility for reviewing protocols, evaluating the biology of the host-vector systems, determining if the DNA to be recombined is well-characterized and free of harmful sequences, inspecting physical facilities, and setting containment levels, has been delegated to the local IBC. Experiments which must be reviewed by the NIH, such as those dealing with cloning of toxin genes, dissemination of recombinant DNA in the environment, and the introduction of antibiotic resistance genes into microorganisms not known to acquire them naturally must be reviewed by RAC and approved by the NIH.

IX. COMMERCIAL DEVELOPMENT OF HUMAN INSULIN, INTERFERON, AND GROWTH HORMONE UNDER THE GUIDELINES

Companies in voluntary compliance with the NIH Guidelines, developed for commercial production human insulin, interferon, and growth hormone using recombinant DNA techniques. During the period from June 1979 to April 1982, a total of 31 proposals involving large-scale operations were reviewed and approved by the RAC and the NIH. Of these proposals, 21 were commercial projects with the ultimate goal of commercially producing human interferon, insulin, or human growth hormone.

The first proposal involving large-scale operations received and reviewed by the NIH dealt with the commercial production of human insulin. This proposal was submitted to the NIH by Eli Lilly and Company in June 1979. In that first proposal, the gene for the insulin A chain was to be cloned in one host-vector system, while the gene for the insulin B chain was to be cloned in a second separate host-vector system. The genes coding for the insulin subunits were chemically synthesized.[16] The molecules produced by the host-vector systems would be "fusion" products of the insulin subunit (either A or B) fused to β-galactosidase.[17] The products would be purified, treated chemically and mixed together to generate active insulin.

Several sets of new procedures had to be devised for the review as this submission was the first requested exception to the prohibition against scale-up, and was submitted under Part VI of the Guidelines. Prior to receiving this request all other RAC business had been conducted in sessions open to the public pursuant to Public Law 92:463, however, proposals involving proprietary information submitted under Part VI had to be reviewed in sessions closed to the public. In addition, those individuals involved in the review assumed responsibilities under 18 U.S.C. 1905, which makes it a "crime for an officer or employee of the

---

* For explanation to footnote see page 164.

Table 4
MODIFICATIONS IN LARGE-SCALE PROCEDURES

Year	Status of Large-Scale Procedures
1976	Large-scale experiments deferred. Experiments of direct societal benefit may be excepted from the deferral provided the experiments are expressly approved by the NIH.
1978	Large-scale procedures prohibited unless the recombinant DNAs are rigorously characterized and the absence of harmful sequences established. Exceptions from the prohibition may be obtained if the experiments are expressly approved by the NIH.
1981	Responsibility delegated to IBCs to review large-scale procedures involving *E. coli* K-12 EK1, *S. cerevisiae*, and *B. subtilis* HV1 host-vector systems.
1982	Responsibility delegated to IBCs to review all large-scale procedures covered by the Guidelines, with the exception of those experiments specifically requiring RAC review and approval (such as those involving the cloning of genes encoding toxins, introduction of antibiotic resistance genes into organisms which would not otherwise acquire this resistance, and dissemination of recombinant DNA containing organisms into the environment.)

United States . . . to publish, divulge, disclose, or make known . . . information which concerns or relates to trade secrets . . . ''

A second class of procedures, those dealing with the manner in which the review would be conducted, were also developed (see Table 3). As part of the review procedure, RAC requested and reviewed information on the physical facilities, in contrast to procedures for review of experiments involving less than 10 ℓ of culture. Another uncertainty was determining appropriate containment conditions. A working group had been established in May 1979 to examine physical containment requirements for large-scale procedures. This group was still evaluating physical containment requirements when the Eli-Lilly proposal was reviewed. Thus, the RAC could not refer to established accepted standards when setting containment conditions. In this circumstance, RAC set containment conditions of ''A P2 facility housing fermentors modified and tested to totally contain the recombinant organisms.'' (The working group subsequently developed standards and these standards were published in the *Federal Register* in April 1980.)*

Approval of the Eli Lilly proposal was recommended by the RAC and the NIH approved the submission in October 1979. Approval was contingent upon acceptance of the condition that ''An observer, designated by the NIH, be permitted to visit the facilities should NIH choose to inspect the site.''[23] This contingency was recommended by the RAC and was imposed on all subsequent Part VI submissions.

Several subsequent proposals involving the commercial production of insulin were received from Eli Lilly and Company, Genentech, Inc., and Cetus Corporation. These proposals involved cloning of a gene coding for proinsulin. NIH review of the earliest proinsulin proposals, submitted by Eli Lilly and Company and Genentech Inc., occurred before the

---

*   In May 1979, the RAC established a working group to examine physical containment requirements for large-scale procedures. At the December 1979 meeting, RAC reviewed a draft proposal from the Large-Scale Working Group on physical containment guidelines for large-scale uses of organisms containing recombinant DNA molecules. Comments from interested institutions and individuals were solicited. In March 1980, the working group reported to RAC that the proposed containment standards for large-scale research and production had been revised in light of comments received on the first draft. The RAC recommended that the revised large-scale standards be published in the *Federal Register* as recommended guidelines. The ''Physical Containment Recommendations for Large-Scale Uses of Organisms Containing Recombinant DNA Molecules,'' which define large-scale physical containment levels called P1-LS, P2-LS, and P3-LS, appeared in the *Federal Register* on April 11, 1980, *Federal Register*, 45, 24968, 1980.

NIH had issued its Physical Containment Recommendations for Large-Scale Uses of Organisms Containing Recombinant DNA Molecules. Containment for the earliest proposals involving proinsulin was therefore "a P2 facility housing fermentors modified and tested to totally contain the recombinant organisms." The Physical Containment Recommendations for Large-Scale Uses of Organisms Containing Recombinant DNA Molecules were published in April 1980, and containment for proposals submitted after that date was set according to standards specified in the Recommendations. The proposals submitted to the NIH involving commercial production of human insulin are summarized in Table 5. As can be seen on the table, all of the proposals involving insulin products specified use of EK1 host-vector systems.

Proposals dealing with human growth hormone were submitted to the NIH by Genentech, Inc. The gene to be cloned would be a combination of chemically synthesized DNA and cDNA. The protein would be directly expressed as active human growth hormone. The earliest proposal dealing with human growth hormone was submitted in November 1979 and was reviewed under a regime similar to that followed for the first insulin submission. Containment for subsequent proposals was set in accordance with the standards specified in the Physical Containment Recommendations for Large-Scale Uses of Organisms Containing Recombinant DNA Molecules.[15] The proposals involving commercial production of human growth hormone are summarized in Table 6.

Proposals dealing with commercial production of human interferon were submitted by several companies: Burns-Biotec Laboratories Inc., Schering-Plough Corporation, Genentech, Inc., Hoffman LaRoche, Inc., and Cetus Corporation. The first proposal for large-scale procedures designed to produce interferon was submitted by Burns-Biotec Laboratories, Inc., a wholly owned subsidiary of Schering-Plough Corporation, in June 1980. As human interferon related sequences fall in several distinct classes.[18,19] the possibilities as to what class of interferon was to be cloned were greater. Thus, the proposals submitted for NIH review were more diverse than those dealing with human insulin or human growth hormone. No attempt to review that diversity is attempted in this article. It can be noted nonetheless that of the protocols submitted for NIH review, only proposals involving the production of interferon specified use of host-vector systems other than *E. coli* EK1. Containment for these procedures was set in accord with the Physical Containment Recommendations for Large-Scale Uses of Organisms Containing Recombinant DNA Molecules. The proposals involving commercial production of human interferon are summarized in Table 7.

Table 5
PROJECTS, SUBMITTED TO THE NIH FOR REVIEW, INVOLVING HUMAN INSULIN

Company	Gene Cloned	Date of Submission	Date of NIH Approval	Date of Federal Register Announcement	Host-Vector	Containment
Eli Lilly	Human Insulin A Chain Human Insulin B Chain	6/8/79	10/5/79	11/30/79	EK1	"in a P2 facility housing fermentors modified and tested to totally contain the recombinant organisms."
Genentech	Human Insulin A Chain Human Insulin B Chain	8/23/79	12/3/79	1/17/80	EK1	"in a P2 facility housing fermentors modified and tested to totally contain the recombinant organisms."
Genentech	Human Insulin A Chain Human Insulin B Chain Human Proinsulin	2/15/80	4/9/80	4/30/80	EK1	"in a P2 facility housing fermentors modified and tested to totally contain the recombinant organisms."
Eli Lilly	Human Proinsulin	2/20/80	4/7/80	4/30/80	EK1	P1—LS
Eli Lilly and Genentech	Human Proinsulin Human Insulin A Chain Human Insulin B Chain	3/6/80	4/9/80	4/30/80	EK1	P2—LS and P1—LS
Cetus	Human Proinsulin	8/12/81	10/21/81	10/30/81	EK1	P1—LS

Table 6
PROJECTS, SUBMITTED TO THE NIH FOR REVIEW, INVOLVING HUMAN GROWTH HORMONE

Company	Gene Cloned	Date of Submission	Date of NIH Approval	Date of Federal Register Announcement	Host-Vector	Containment
Genentech	Human Growth Hormone	11/12/79	4/1/80	4/30/80	EK1	"in a P2 facility housing fermentors modified and tested to totally contain the recombinant organisms."
Genentech	Human Growth Hormone	5/12/80	7/22/80	8/21/80	EK1	P1—LS

Table 7
PROJECTS, SUBMITTED TO THE NIH FOR REVIEW, INVOLVING HUMAN INTERFERON

Company	Gene Cloned	Date of Submission	Date of NIH Approval	Date of Federal Register Announcement	Host-Vector	Containment
Burns-Biotec Laboratories	Human Leukocyte Interferon	5/14/80	7/22/80	8/21/80	EK1	P1—LS
Genentech	Human Leukocyte Interferon Human Fibroblast Interferon	5/13/80	7/22/80	8/21/80	EK1	P1—LS
Genentech	Human Leukocyte Interferon Human Fibroblast Interferon	9/4/80	11/4/80	11/21/80	EK1	P1—LS
Burns-Biotec Laboratories	Human Leukocyte Interferon	9/12/80	11/4/80	11/21/80	EK1	P1—LS

Table 7 (continued)
PROJECTS, SUBMITTED TO THE NIH FOR REVIEW, INVOLVING HUMAN INTERFERON

Company	Gene Cloned	Date of Submission	Date of NIH Approval	Date of Federal Register Announcement	Host-Vector	Containment
Genentech	Human Leukocyte Interferon	12/5/80	2/2/81	3/12/81	EK1	P1—LS
Hoffman La-Roche	Human Leukocyte Interferon Human Fibroblast Interferon	12/9/80	2/2/81	3/12/81	EK1	P1—LS
Burns-Biotec Laboratories	Human Leukocyte Interferon	12/10/80	2/20/81	3/12/81	EK1	P1—LS
Schering-Plough	Human Leukocyte Interferon	12/10/80	2/20/81	3/12/81	EK1	P1—LS
Cetus	Human Fibroblast -Interferons	4/2/81	6/16/81	7/1/81	EK1 HV1 *B. subtilis*	P1—LS
Genentech	Human Leukocyte Interferons A and D	8/5/81	10/21/81	10/30/81	*S. cerevisiae*	P1—LS
Schering-Plough	Human Leukocyte Interferons	8/4/81	10/21/81	10/30/81	EK1	P1—LS
Cetus	Human alpha-1 Interferons	8/12/81	10/21/81	10/30/81	EK1	P1—LS

REFERENCES

1. **Singer, M. and Soll, D.,** Guidelines for DNA hybrid molecules, *Science,* 181, 1114, 1973.
2. Recombinant DNA Research Guidelines, *Federal Register,* 41, 27902, 1976.
3. Recombinant DNA Research Proposed Revised Guidelines, *Federal Register,* 43, 33042, 1978.
4. Guidelines for Research Involving Recombinant DNA Molecules, *Federal Register,* 43, 60108, 1978.
5. Guidelines for Research Involving Recombinant DNA Molecules, *Federal Register,* 45, 6724, 1980.
6. Guidelines for Research Involving Recombinant DNA Molecules, November, *Federal Register,* 45, 77384, 1980.
7. Guidelines for Research Involving Recombinant DNA Molecules, June, *Federal Register,* 46, 34462, 1981.
8. Recombinant DNA Research; Proposed Actions Under Guidelines, *Federal Register,* 46, 17994, 1981.
9. Recombinant DNA Advisory Committee, Minutes of the Meeting, September 10—11, 1981, 3 and 17, Department of Health and Human Services, Bethesda, Md.
10. Evaluation of the Risks Associated with Recombinant DNA Research, *Federal Register,* 46, 59384, 1981. and *Recombinant DNA Tech. Bull.,* 4, 4, 166, 1981.
11. Recombinant DNA Research, Proposed Actions Under the Guidelines, *Federal Register,* 46, 59368, 1981.
12. Recombinant DNA Research, Proposed Actions under the Guidelines, *Federal Register,* 46, 59734, 1981.
13. Guidelines for Research Involving Recombinant DNA molecules, *Federal Register,* 47, 17180, 1982.
14. Guidelines for Research Involving Recombinant DNA Molecules, *Federal Register,* 47, 38048, 1982.
15. Recombinant DNA Research; Actions Under Guidelines, *Federal Register,* 46, 53980, 1981.
16. **Goeddel, D. V., Kleid, D. G., Bolivar, F., Heyneker, H. L., Yansura, D. G., Crea, R., Hirose, T., Kraszewski, A., Itakura, K., and Riggs, A. D.,** Expression in *Escherichia coli* of chemically synthesized genes for human insulin, *Proc. Natl. Acad. Sci.,* USA, 76, 106, 1979.
17. **Crea, R., Krazsewski, A., Hirose, T., and Itakura, K.,** Chemical synthesis of genes for human insulin, *Proc. Natl. Acad. Sci.,* U.S.A., 75, 5765, 1979.
18. **Nagata, S., Mantei, N., and Weissman, C.,** The structure of one of eight or more distinct chromosomal genes for human interferon-α, *Nature,* 287, 401, 1980.
19. **Allen, G. and Fantes, K. H.,** A family of structural genes for human lymphoblastoid (leukocyte-type) interferon, *Nature,* 287, 408, 1980.
20. **Berg, P., et al.,** Potential Hazards of Recombinant DNA molecules, *Science,* 185, 303, 1974.
21. Recombinant DNA Advisory Committee, Minutes of the Meeting, April 23-24, 1981, 14, Department of Health and Human Services, Public Health Service, Bethesda, Md.
22. Guidelines for Research Involving Recombinant DNA Molecules, *Federal Register,* 47, 17187, 1982.
23. Recombinant DNA Research; Proposed Guidelines and Actions, *Federal Register,* 44, 69251, 1979.

Chapter 10A

APPENDIX

VIRAL VECTORS AND THE NIH GUIDELINES

Stanley Barban

TABLE OF CONTENTS

I. HISTORY

The containment conditions specified by the first NIH Recombinant DNA Guidelines promulgated in 1976 for experiments with animal viruses were extremely stringent. The levels of physical containment for cloning of animal virus genes in the *E. coli* K-12 host-vector system ranged from P3 to P4. For biological containment the use of the specially enfeebled strains of *E. coli*, the EK2 systems, were mandated for many experiments. Further, the employment of certain animal virus vectors such as the defective papova viruses, SV40 and polyoma, called for either P3 or P4 physical containment conditions for cloning in mammalian cells. The defective virus vectors lacked substantial sections of their genomes, generally the late regions involving capsid formation, which rendered them incapable of self-replication. Non-permissive cell lines which did not permit infectious virus particles to be produced were required for transfection experiments with the viral recombinants. The rescue of infectious virus particles from cell cultures with helper virus, especially with SV40, required P4 physical containment. The basis for these restrictive conditions were questions raised concerning possible risks developed from the viral recombinants. The major potential risk envisioned was the possibility that the defective recombinant virus molecule could be encapsidated and be able to replicate and propagate as a virus with unique properties. Could the recombinant virus manifest a significant change in its mechanism or mode of pathogenicity? Could the recombinant virus possess a selective advantage in the host by virtue of greater stability and host range specificity or contain cellular or viral oncogene segments? It was apparent that the stringent physical constraints of the early Guidelines severely inhibited basic research studies in the development of new transducing vectors for mammalian cells.

Containment conditions for the use of recombinant viral vectors have undergone a drastic evolution and deregulation over the past years. The rationale for these changes was the expansion of our knowledge based on conclusions of several workshops of experts in the field of virology, the results of certain risk assessment experiments involving derivatives of *E. coli* K-12 carrying a complete viral genome, the polyoma risk assessment experiments, and new information about the molecular biology of viral gene expression. The considerations and recommendations of the participants of the joint U.S. -EMBO Workshop held in Ascot, England, January 27 to 29, 1978,[1] provided important guidance and reference for a major revision in the Guidelines. In addition, the suggestions of a working group on eukaryotic viral vectors of the Recombinant DNA Advisory Committee on April 13, 1980, at the American Society for Microbiology meeting in Miami Beach, Florida were also a basis for a significant change in the Guidelines.

The scientists assembled at the Ascot meeting to assess the risks for recombinant DNA experiments involving the genomes of animal, plant, and insect viruses were experts in public health, medical and clinical diagnostic virology, and biochemical virology. The participants of that meeting discussed extensively the state of the art in safely working with various animal viruses in the laboratory. Special emphasis was given to those viruses that could potentially be employed as transducing vectors in mammalian cells. Various animal viruses were categorized on the basis of their known potential for the severity of human disease they cause, laboratory infections, and possible effect of spread of the virus to the community or environment. These experts, after careful discussions of the possible risks associated with cloning eukaryotic viral DNA sequences in the *E. coli* K-12 host-vector system and with the use of various eukaryotic viruses as cloning vectors in animal, plant, and insect systems, concluded with several recommendations. A recommendation was put forth that P2 physical containment conditions in conjunction with an EK1 host-vector system was a minimum level for recombinant DNA experiments involving eukaryotic viral DNA inserts consisting of genomic or subgenomic fragments. They stressed, however, that if the wild type virus itself required higher levels of physical containment for handling, then more stringent containment conditions should be employed when working with the recombinant DNA virus particles.

The polyoma cloning experiments of Israel, Chan, Garon, Martin, and Rowe[2,3] were designed as a model system for evaluating certain postulated biohazards associated with recombinant DNA technology with viral DNAs. The important questions to be addressed were whether viral DNA, covalently linked to a vector, could be transferred from within a bacterial cell, *E. coli* K-12 or derivatives, to cultured cells, or under in vivo conditions to initiate viral infection or induce tumors in animals. The results of these risk assessment experiments demonstrated that recombinant plasmids containing potentially infectious genomic polyoma DNA failed to induce polyoma infection following inoculation or feeding of large doses to permissive mice. Comparable results employing cell cultures were also reported by Fried et al.[4] Recombinant plasmid and phage vectors carrying monomeric polyoma inserts were found to be non-infectious. Extension of the polyoma virus risk assessment experiments by Israel, Chan, Martin, and Rowe to test tumor induction of the virus recombinants was also reported.[5] When baby hamsters were inoculated with an *E. coli* K-12, EK2 system, carrying the complete genome of polyoma virus in a recombinant plasmid, no tumors were observed. These findings served as a support for major changes which were subsequently made in the lowering of containment for recombinant experiments with both animal and plant viruses during the constant evolution of the Guidelines over the past years.

At the American Society for Microbiology meeting in Miami Beach, Florida, April 13, 1980, a working group of distinguished virologists considered the appropriate containment conditions for the rescue of viral recombinants containing less than $^2/_3$ of the genome of any eukaryotic virus. Their recommendations were adopted by the Recombinant DNA Advisory Committee and subsequently incorporated in the revisions of the Guidelines promulgated by the NIH on November 21, 1980.

Tables 1 and 2 illustrate the radical changes in the levels of containment requirements for using eukaryotic virus vectors (Table 1) and for cloning of viral DNA (Table 2) in different revisions of the NIH Guidelines.

II. CURRENT STATUS OF THE GUIDELINES FOR THE USE OF RECOMBINANT DNA MOLECULES

Although the Guidelines have undergone further evolution since November 1980, including major revisions on July 1, 1981, April 21, 1982, and August 27, 1982, the final chapter has not been written. In the interim, following these radical changes in containment conditions for cloning viral DNA and use of viral vectors, major advances have been made in the development of new cloning vehicles for introducing genetic sequences into animal cells and with it there has been a rapid expansion of our knowledge on the molecular analysis of oncogenic viruses, possible mechanisms of gene regulation and cell transformation. The development and application of recombinant DNA technology has indeed opened a new era of scientific discovery.

Table 1

COMPARISON OF THE CONTAINMENT LEVELS FOR USE OF CERTAIN VIRAL VECTORS WITH REVISIONS OF THE GUIDELINES

		Physical containment levels in guidelines		
Vector DNA	Viral DNA inserts	June 1976	December 1978	November 1980
SV40 (deleted genome)	DNA-non-pathogen[a]	P4	P2	P1[b]
Polyoma (deleted genome)	DNA from non-pathogen	P4	P2	P1
		or		
		P3	P2	P1
Human Adenovirus 2, 5 (defective)	Subgenomic	—	P3/P2	P1
Retroviruses (deleted genome)	Subgenomic	—	CBC[c]	P2

[a]   DNA from any non-pathogenic organism or Class 1 virus.
[b]   Recombinant DNA Virus molecules containing no more than $^2/_3$ of the genome of any eukaryotic virus [all viruses from a single Family being considered identical]. The host cells were not to contain helper virus for the specific Families of defective viruses being used. The DNA inserts could contain less than $^2/_3$ of genomes of viruses from more than one Family.
[c]   CBC denotes review by the NIH on a case-by-case basis.

Table 2

COMPARISON OF THE CONTAINMENT LEVELS FOR CLONING OF VIRAL DNA IN *E. COLI* K-12 HOST-VECTOR SYSTEM WITH REVISIONS OF THE GUIDELINES

	Containment Levels in Guidelines		
	June 1976	December 1978	November 1980
Virus Class	Physical & Biological	Physical & Biological	Physical & Biological
Animal Viruses			
DNA Transforming Viruses	P4 + EK2	P2 + EK2	P1 + HV1[a]
(When clones or subgenomic	or		
segments—free of harmful	P3 + EK2	P2 + EK1CV[b]	
regions)			
RNA Retroviruses	P3 + EK1	P2 + EK1CV	P1 + HV1

[a]   HV1 denotes a certified host-vector system which provides a moderate level of containment, i.e., *Saccharomyces cerevisiae, B. subtilis.*
[b]   EK1CV means the use of an EK1 host and a vector certified for use in an EK2 system.

REFERENCES

1. **Martin, M. A., Rowe, W. P., and Tooze, J.,** Report of US-EMBO workshop to assess risk for recombinant experiments involving the genomes of animal, plant, and insect viruses, *Recombinant DNA Technical Bulletin,* 1, 24, 1978.
2. **Israel, M. A., Chan, H. W., Rowe, W. P., and Martin, M. A.,** Molecular cloning of polyoma virus DNA in *Escherichia coli,* Plasmid vector system, *Science, 203,* 883, 1979.
3. **Chan, H. W., Israel, M. A., Garon, C. F., Rowe, W. P., and Martin, M. A.,** Molecular cloning of polyoma virus DNA in *E. coli,* Lambda phage vector system, *Science,* 203, 887, 1979.
4. **Fried, M., Klein, B., Murray, K., Greenaway, P., Tooze, J., Boll, W., and Weissman, C.,** Infectivity in mouse fibroblasts of polyoma DNA integrated into plasmid pBR322 or lambdoid phage DNA, *Nature,* 279, 811, 1979.
5. **Israel, M. A., Chan, H. W., Martin, M. A., and Rowe, W. P.,** Molecular cloning of polyoma virus DNA in *Escherichia coli,* Oncogenicity testing in hamsters, *Science,* 205, 1140, 1979.

Chapter 11

FDA'S ROLE IN APPROVAL AND REGULATION OF RECOMBINANT DNA DRUGS

Wanda deVlaminck*

TABLE OF CONTENTS

*   This chapter was written while Ms. deVlaminck was Director, Regulatory Affairs — Interferon Shell Oil Company, Berkeley, California.

I. INTRODUCTION

Biotechnology, which uses microorganisms to produce drug products, is certainly not a revolution in the pharmaceutical industry; for decades, microorganisms have been used in the manufacture of antibiotics. Over 12 years ago, L-asparaginase, a protein from *E. coli*, was approved by the Food and Drug Administration (FDA) for use in patients with acute lymphatic leukemia.[1] The revolution in biotechnology is mainly the result of the introduction of genetically engineered microorganisms into a variety of industrial processes, resulting in recombinant DNA drug products.

Many exciting scientific discoveries resulting from biotechnology will have ultimate meaning when practical application results in the production of safe and effective drugs. As the biotechnology revolution gains momentum and these new discoveries follow the arduous path towards becoming approved drugs, the FDA will play the primary and official role in establishing the regulatory requirements for these new drugs. There will be continuing pressure from opposing forces to approve new drugs in shorter periods of time with no compromise to the public health. The important questions now being asked are, how will the FDA meet the challenge of regulating recombinant DNA drugs, what new regulations will be promulgated, and how will existing regulations be applied?

There are numerous indications that the FDA is prepared to meet this challenge by increasing its scientific capabilities and interacting closely with industry and academia. The regulatory requirements seem to be evolving in harmony with the biotechnology revolution and the future for new drug development in the United States looks quite promising.

II. THE FDA'S CHANGING ROLE

Technological revolutions of the past that relate to drug discoveries and development have often resulted in a backlash of regulatory activities. In some cases this backlash has resulted in meaningful revisions or additions to existing laws governing drugs. For instance, the thalidomide tragedy in the early 1960s precipitated the enactment of regulations applicable to the testing of investigational drugs.

In other cases, emotional public reaction has overriden scientific rationale. For example, after the thalidomide tragedy, a new approach was initiated for preclinical animal studies that required large numbers of animals to be given large amounts of a drug for an extended period of time.[2] This approach became a routine requirement with little or no scientific validity, and may have contributed to the demise of some potentially important new drugs.

A more enlightened approach to drug regulation seems to be presently taking place. There has been a conscious effort by the FDA to closely follow the development of recombinant DNA technology and begin putting in place a staff of experts who would be able to establish the scientific base necessary to evaluate, test, and approve the "new technology" drugs. This began to occur over 5 years ago when somatostatin become the first human peptide to be synthesized in a bacterial cell.[3]

The federal government has recognized that the FDA will have to interact with industry and academia in order to deal effectively with regulatory issues regarding recombinant DNA drugs. As a result of this awareness, FDA sponsored workshops and seminars have been held in the past and undoubtedly will be held in the future to discuss issues regarding the safety and efficacy of recombinant DNA products. Such interactions enforce the concept that the role of the FDA is not based on scientific constants that do not change. Rather, the FDA is looking to the responsible biomedical community and the pharmaceutical industry to help establish requirements for recombinant DNA drugs and drugs in general.

The regulatory process for drug approval has come under increasing scrutiny in the last several years. The average time that it takes to obtain an approval for a new drug starting with an Investigational New Drug (IND) filing was $3^1/_2$ years in 1967 compared to approximately 10 years in 1982.[2] Due to this increased amount of time, the United States has experienced "drug lag" compared to other technologically advanced countries.

According to the final report issued by the Commission on the Federal Drug Approval Process dated March 31, 1982, "The numbers of new chemical entities entering clinical research and those ultimately approved for marketing have declined dramatically in recent years. Although the reasons for this decline are many and diverse, it is generally perceived that regulatory demands and delays within the United States drug review system are significant factors."[4] It is interesting to note that the former Commissioner of the FDA, Dr. Arthur Hall Hays, Jr., has acknowledged that there is a drug lag in the United States and that the FDA is at least partially responsible for it.[5]

Consistent with the acknowledgement of a drug lag in this country and the liberal attitude of the present Reagan administration towards regulations in general, major revisions in the drug approval process have been proposed. If adopted, these revisions could reduce the total time required for review of a New Drug Application (NDA) by 6 months.[6] Revisions in the NDA review process will have a direct effect on approvals for recombinant DNA drugs. However, the Investigational New Drug (IND) process is the most time consuming and complex part of drug development and has the most impact on recombinant DNA drugs at the present time. A proposal to revise the requirements for IND's is scheduled for late 1982.

The reorganization of the FDA-Bureau of Drugs and the Bureau of Biologics will help to develop the most efficient and consistent approach towards the regulation of bacteria-derived drugs such as insulin and growth hormone, and the bacteria-derived biologicals such as the interferons; alpha, beta, and gamma.

The Bureaus of Drugs and Biologics have been merged to form the National Center for Drugs and Biologics under the FDA. The new center is organized into three groups under the director's office which is staffed by management personnel from the original bureaus:

- Office of New Drug Evaluation — This office consists of 6 divisions to review and approve new drug applications.
- Office of Drugs — This office consists of 9 divisions which include compliance activities, adverse drug reaction program, and the development of standards for Over-The-Counter (OTC) drugs.
- Office of Biologics — This office consists of seven divisions which regulate and conduct research on biological products.

Prior to the aforementioned reorganization, a Recombinant DNA Advisory Committee was formed within the FDA. The function of the committee is to facilitate an organized approach to information exchange, determine policies by way of consensus decisions, and determine jurisdiction among FDA offices for recombinant DNA drugs.

There is no doubt that the FDA is making every effort to meet the challenge of regulating recombinant DNA drugs and at the same time respond to the diverse and often conflicting demands of the Federal Government, the medical community, the pharmaceutical industry, and the public as well as the consumer advocate groups.

III. NEW REGULATIONS FOR RECOMBINANT DNA DRUGS

Recombinant DNA technology used for the manufacture of human proteins and polypeptides represents a more sophisitcated and technical approach to the classic fermentation technology developed to manufacture antibiotics and steroids. Advances in genetics, which took place in the mid-1940s,[3] initiated the development of fermentation technology as we know and use it today. Induced mutations and the concept of exchanging genetic material between organisms to amplify certain desired traits has been the basis for developing industrial scale fermentation processes. Recombinant DNA technology takes this process one step further by adding the gene for the desired product to a microorganisms rather than being restricted to only those products that result from an organism's naturally occurring DNA.

Fermentation technology has genetically manipulated microorganisms to become more efficient producers of their natural metabolites. However, recombinant DNA technology has genetically manipulated microorganisms to produce products such as polypeptides and proteins that are totally foreign to them.

Even though the pharmaceutical industry has been using fermentation for years to produce antibiotics, there has not been much experience with the fermentation of proteins, especially proteins produced by recombinant DNA microorganisms. Consequently, the challenging regulatory issues will relate more to the products resulting from the recombinant DNA technology than the process itself. Good Laboratory Practice regulations[14] and current Good Manufacturing Practice regulations[15] provide an existing framework for regulation of the preclinical testing and production and quality control. These regulations have been in effect for some time and are followed extensively by the pharmaceutical industry.

It is too soon to know what specific new regulations will be promulgated for recombinant DNA drugs. However, it can be assumed that some specific concerns relating to safety, efficacy, purity, and potency will be expressed by the FDA. The FDA has taken the position that drugs produced by recombinant DNA technology will be considered new drugs since they have not been generally recognized by qualified scientific experts as safe and effective. An IND application and an NDA will be required even if the product is identical to an existing product or a natural human substance. Specific requirements relating to a particular product class will be regulated on a product-by-product basis by the appropriate office of the FDA.[7]

Microorganisms that contain eukaryotic genes may be unstable and vulnerable to mutation which could result in the production of mutant polypeptides. Therefore, the DNA sequence in the recombinant microorganism will probably be requested in order to provide added assurance as to the identity and purity of the final recombinant DNA product. Use of sensitive analytical procedures such as HPLC tryptic mapping will be encouraged in order to monitor in-process material as well as the final product, and to validate the stability of the plasmids.

In addition to producing a specific hormone or structural protein, a recombinant microorganism will also produce significant levels of its own bacterial proteins or non-protein materials. Since recombinant microorganisms currently being used for production do not secrete the desired product, microbial polypeptides, lipids, carbohydrates, and nucleic acids are harvested along with the desired product when the microbial cell wall is lysed. Therefore it will be essential to develop purification processes that are effective and to develop sensitive analytical techniques for evaluation of purity.

Immunologic potential, pharmacokinetics, and bioactivity of the recombinant DNA drug product are undoubtedly related to the amino acid sequence and/or the structure of the molecule. Implications as to safety and efficacy that relate to differences between the native and recombinant molecules can be most accurately assessed as a result of clinical trials. Therefore, sensitive assays and scientifically valid protocols must be developed in order to obtain meaningful results from clinical studies. Even when the amino acid sequence of the recombinant DNA protein is exactly the same as the native protein, clinical questions regarding efficacy must be addressed.

Biosynthetic human insulin (BHI) produced in *E. coli* has an amino acid sequence and structure identical to that of native human insulin and shows similar biological activities in animal and human studies when compared to the native material.[8] However most of the clinical experience with insulins has resulted from the use of animal derived insulins; the difference between human and porcine insulin is 1 amino acid, and the difference between human and bovine insulin is 3 amino acids. These differences account for allergic reactions and antibody response in approximately 5% of the diabetics who must take insulins presently on the market.[9]

In a recent double-blind crossover study conducted in Great Britain,[8] diabetic patients received 6-week periods of treatment with either porcine or bovine insulin and then received an additional 6-week treatment period of BHI. The results of the study indicate that BHI is a potent insulin preparation with no obvious side effects, but that the BHI has slightly different pharmacokinetic properties which resulted in higher glucose levels in some patients.[8] Studies done in non-diabetic subjects showed that BHI is absorbed more rapidly and therefore probably excreted or inactivated more rapidly than purified porcine insulin.[8]

The purity of the BHI would suggest that antibody formation to either the insulin or contaminants will not present a problem. However, additional clinical studies must be done to assess behavior of pre-existing insulin antibodies in diabetics treated with animal insulin preparations.[8]

Standard potency assays for conventional products of human growth hormone and insulins have been in use for many years and these assays can be applied to recombinant DNA human growth hormone and insulin. In the case of interferons, no acceptable standard potency assay exists.[10] Numerous types of bioassays are being used to identify and measure interferon activity. Variables in the bioassay include the cell substrate, endpoint, and the challenge virus, all of which can independently or cumulatively contribute to varying results. With Phase I and II clinical studies in progress it is essential that a standard bioassay be developed and designated as such so that the clinical investigators and the FDA can accurately assess efficacy and dose response.

International interferon reference preparations, with unitage expressed in international units, have been prepared by the FDA — Office of Biologics and the National Institutes of Health as a first step towards bioassay standardization. These international reference preparations can be used to calibrate interferon assays and evaluate in-house reference preparations. Once the in-house reference preparation is calibrated against the international reference, it can be used routinely and test results can be expressed in international units.

IV. HISTORY OF DRUG REGULATION

While the FDA is making efforts to reevaluate existing drug regulations in order to ultimately expedite approvals for all new drugs, including the recombinant DNA drugs, the fact is that most of the existing regulations are here to stay and, in most cases, for good reason.

The scientific accomplishments in the research lab are taking a fast track towards becoming new drugs, so more scientists involved with the development of recombinant DNA drugs are finding it necessary to increase their awareness of existing drug regulations. With a working knowledge of the regulations, appropriate decisions and choices in the research phase of drug development can be made which may have major significance when an IND or an NDA is submitted to the FDA.

Seventy-six years have passed since the enactment of the first Federal Food and Drug Act.[5,11] Signed by Theodore Roosevelt on June 30, 1906,[11] this Act was mainly the result of industrialization and urbanization of food production and distribution. Mass production and processing of food required new techniques for storage and shipment. Preservatives such as borax and formaldehyde were widely used with no concern for safety for consumers. Drugs too, were mass produced and could be purchased directly from a pharmacist without a doctor's prescription. Inappropriately labelled drugs containing alcohol and narcotics were widely distributed.[11]

The new law banned from interstate commerce adulterated or misbranded food or drugs and made manufacture of such items unlawful, but even though the law required adequate labelling, it did not ban hazardous ingredients as long as the label accurately stated their presence.

It wasn't until 1912 that legislation was enacted to regulate therapeutic claims by passing the Sherley Amendment.[11] This amendment didn't have much enforcement clout, since the burden was on the government to prove intent to deceive.

The FDA became an official government agency when the Bureau of Chemistry was shifted to the Food, Drug and Insecticide Administration under the Department of Agriculture in 1927, and renamed the Food and Drug Administration in 1931.[11]

Laws governing food and drugs generally were accepted by the public with no reluctance. Not until Franklin D. Roosevelt's administration and the depression did proposals for additional regulation come under attack. Revision of the Food and Drug Act, a mandate of the Roosevelt administration, met with strong opposition from the private sector. Specifically, industry had concerns about any government control due to the potential economic effects on an already gravely depressed economy.

The "Elixir Sulfanilamide" disaster on June 25, 1938, turned the tide towards an expanded role of drug control by the federal government. A drug manufacturer in Tennessee used diethylene glycol to produce a sulfa drug in suspension,[11] but failed to conduct any safety studies on this product, and it resulted in over 100 known deaths. Federal, state, and local drug authorities acted quickly when it became apparent that this elixir was the cause of numerous deaths. Approximately 99% of the lot was seized prior to distribution, circumventing possibly thousands of deaths.

After this incident, public pressure supported comprehensive legislation that would expand the role of government in regulation and approval of drugs. The Food, Drug and Cosmetic Act of 1938 became the basic federal statute governing the manufacture, testing, approval, and distribution of drugs as well as foods and cosmetics. Drug manufacturers were required to show evidence of safety before a drug could be marketed. Non-prescription drugs, referred to as "Over-The-Counter" (OTC) drugs were required to be marketed with information indicating the appropriate use as well as warnings regarding misuse. However, the law did not dictate which drugs were to be sold as prescription or non-prescription drugs; the effect was that nearly all new drugs were sold by prescription. Since only physicians could prescribe drugs, they became instrumental in deciding what drugs would be used by the public.[13]

It wasn't until 1962 that drug regulations were revised to the extent that we know them today. One of the most significant additions to the laws governing new drug approval came as a result of the thalidomide tragedy. Even though thalidomide was not an approved drug in the United States, the drug had been used extensively in investigational studies and birth defects had resulted. As a result of the 1962 Drug Amendments, clinical studies for new drugs came under the jurisdiction of the FDA. The FDA developed an extensive set of requirements for conducting clinical studies in support of new drug approvals.

In addition to IND requirements, the 1962 Drug Amendments expanded FDA's authority in other areas of drug regulation. Substantial evidence was required to show that new drugs were effective in addition to being safe. Even labeling indicating efficacy was to be reviewed by the FDA for false and misleading statements. Manufacturers were required to report adverse drug reactions to the FDA. Authority to inspect drug manufacturing facilities was expanded to include the Good Manufacturing Regulations and provisions were developed to allow the FDA to remove a drug from the market if regulations were not met.

V. OVERVIEW OF EXISTING REGULATIONS

A typical sequence of events occurs when a potential drug product goes from the research and development laboratory into clinical Phase I testing. In the case of recombinant DNA drugs, some of these events may be abbreviated or expanded, and since each new recombinant DNA Drug must be evaluated on a product-by-product basis, it is difficult to predict a precise sequence of events.

The Federal Food, Drug and Cosmetic Act authorizes the FDA to regulate the testing of new drugs, leaving the details up to the FDA. The FDA has promulgated specific regulations that provide an exemption from the requirement for a New Drug Application (NDA) for drugs intended only for investigational use. This "Notice of Claimed Investigational Exemption for a New Drug" is commonly referred to as an IND. The major components of an IND are as follows:

1. Detailed description of the planned clinical investigation including an outline of future phases of the study
2. Description of the drug, including the name, components, composition, and raw material sources
3. Route of administration
4. Information resulting from preclinical studies as well as prior clinical studies
5. Description of the manufacturing process
6. Clinical brochure which consists of informational materials to be supplied to each investigator
7. Description of the training and experience of each investigator
8. Institutional Review Board approval
9. Informed consent form
10. A statement that all preclinical studies were done in compliance with the Good Laboratory Practice Regulations
11. Environmental Impact Analysis Report, if required by the FDA

Prior to filing an IND, it is often beneficial to begin a dialogue with the appropriate office at the FDA. Often a Pre-IND Meeting will accomplish this and set the stage for preliminary information exchange. Proposed preclinical animal testing as well as the proposed Clinical Phase I Study can be discussed. General descriptions of the process and product characterization will help to get initial reactions from the FDA personnel who will be instrumental in the review process.

As the production process becomes more defined, documentation suitable for lab notebooks is replaced by controlled documents in the form of batch production records and approved record forms, since the current Good Manufacturing Practice Regulations also apply to drugs still in investigational stages. These records help to standardize the process because each step is listed and must be verified by the operator. At key steps in the process, product characteristics can be evaluated and compared based on accurate and accessible records.

Preliminary animal studies can provide important data required for the development of preclinical animal studies that will be used to support the IND. The questions to be answered by doing animal studies depend on the product itself, as well as components introduced as part of the process or formulation. Chronic studies in animals using human protein products present difficulties regarding interpretations of results due to the potential reactions of the animals to foreign proteins. Preliminary animal studies provide a setting whereby compliance with GLP regulations can be evaluated before embarking on large scale, costly, labor intensive studies.

As results from animal studies are evaluated, final plans for the clinical studies usually commence. The investigators who will conduct the clinical studies and the manufacturer of the investigational drug discuss the clinical protocol study design and arrive at a consensus regarding details of the study.

Biological and chemical tests are developed for product characterization and, as they become routine test procedures, are turned over to the quality assurance laboratory. With reproducible, sensitive tests, a stability testing program can be initiated. Data resulting from stability studies can be periodically tabulated for evaluation.

Throughout the development process, it is important to maintain routine contact with the regulatory agency. Samples of the product can be submitted for testing in order to get comparative data and detect apparent discrepancies. Early sample submission allows the agency to begin working with the product to develop specific tests or incorporate the product into existing test systems. This is particularly true of the Office of Biologics, since they are often involved in areas of research common to a variety of products that come under their review.

Once the IND is submitted to the FDA, a period of 30 days must elapse before actually initiating clinical studies. A waiver may be granted by the FDA to reduce the 30 day waiting period if extenuating circumstances make it necessary and appropriate. If the FDA takes no action within the 30 day period, the Phase I clinical study may commence. In most cases, enough prior communication has taken place between the regulatory agency and the manufacturer so that unanticipated events will not delay the initiation of the clinical study. The FDA will not disclose the existence of an IND unless prior public disclosure has taken place.

In addition to the requirement for an IND, other requirements must be met for conducting clinical studies, and the FDA may terminate a study if these are not met:

1. Investigational drugs must have the following statement on the label: "Limited by Federal (or US) Law to investigational use."
2. Complete and accurate records must be maintained on drug accountability.
3. A clinical investigation must be monitored by the sponsor of the new drug.
4. The investigational drug may not be sold or promoted as an approved drug. Under certain conditions, the investigational drug may be sold to investigators.
5. Clinical studies must be conducted in a timely manner.
6. When all clinical studies have been completed and indicate that a drug is safe and effective, an NDA must be filed or within 60 days of a request by the Commissioner.
7. No test marketing of an investigational drug is allowed prior to approval.

The purpose for conducting clinical studies is to show that a new drug is safe and effective and meets FDA requirements for approval. When a New Drug Application (NDA) is submitted to the FDA, it must include substantial evidence of safety and efficacy. The FDA defines substantial evidence as adequate and well-controlled investigations conducted by experts qualified by scientific training and experience.

Efficacy does not necessarily mean that a drug has the capacity to cure an illness or clinical conditions. Prolonged life, reduced pain, or improved physical condition can all be used as criteria for judging the efficacy of a new drug.

Safety is judged by the therapeutic benefit compared to the risk of taking the new drug. Clinical studies provide the basis for making such a judgement. Obviously the assessment of benefit/risk of a drug in terminally ill patients is different from that used in the general population to reduce a non-life-threatening condition. Benefit/risk assessments become more complex when a new drug is compared to an approved drug on the market. Carefully designed and controlled clinical studies that address both scientific and ethical questions and concerns must be conducted in order to provide clinically significant information.

VI. CONCLUSION

The pharmaceutical industry and the FDA seem to be approaching the age of recombinant DNA drugs with a similar degree of cautious optimism. The adversary relationships of the past have given way to collaborative interactions which will extend scientific foresight into the unknowns of the future.

Once the recombinant DNA drugs are approved by the FDA, post-marketing surveillance will become an increasingly important means to evaluate their long-term effects. The medical community, industry, and the FDA will share equally in the responsibility of developing and/or conducting such post-marketing studies.

The medical community will have the responsibility to see that recombinant DNA drugs are used with discretion in the most relevant clinical settings. Limited indications for use of approved drugs can be expanded by clinicians within the definition of the practice of medicine. Therefore, it is essential that the medical community rely on the results of adequate and well-controlled clinical studies before extending indications for use.

Cautious optimism may seem too restrained and conservative amidst the flurry and excitement of the biotechnology revolution. There are no short-cuts to replace experience. Good science must provide the foundation on which to build experience if humanity is to derive maximum benefits from this "Cinderella Science" called recombinant DNA technology.

REFERENCES

1. **Grinnan, E. L.**, L-Asparaginase: a case study of an *E. coli* fermentation product, in *Insulins, Growth Hormone, and Recombinant DNA Technology,* Gueriguian, J. L., Ed., Raven Press, New York, 1981, 99.
2. **Weatherall, M.**, An end to the search for new drugs?, *Nature,* 296, 387, 1982.
3. **Aharonowitz, Y. and Cohen, G.**, The microbiological production of pharmaceuticals, *Sci. Am.,* 245, 141, 1981.
4. **McMahon, F.**, Commission of the Federal Drug Approval Process, Committee on Science and Technology, Congressional Committee, Final Report, March 31, 1982.
5. **Ballin, J. C.**, Regulation and development of new drugs, *JAMA,* 247, 2995, 1982.
6. **Hanson, D.**, Plan to speed up drug review approved, *Chem. Eng. News,* 60 (28), 18, 1982.
7. **Miller, H. I.**, The role of the FDA in regulating products of recombinant DNA technology, in *Proceedings 1981 Battelle Conference on Genetic Engineering,* Vol. 2, Keenberg, M., Ed., Battelle Seminars and Studies Program, Seattle, Wash., 1981, 25.
8. **Clark, A. J. L., Knight, G., Wiles, P. G., Keen, H., Ward, J. D., Cauldwell, J. N., Adeniyi-Jones, R. O., Leiper, J. M., Jones, R. H., MacCuish, A. G., Watkins, P. J., Glynne, A., and Scotton, J. B.**, Biosynthetic human insulin in the treatment of diabetes, a double-blind crossover trial in established diabetic patients, *Lancet,* 2, 354, 1982.
9. **Evans, H. J.**, Biotechnology and medicine, *Health Bull.,* 40, 34, 1982.
10. **Grossberg, S. E.**, On reporting interferon research, *Ann. Intern. Med.,* 95, 115, 1981.
11. **Hayes, A. H.**, Food and drug regulation after 75 years, *JAMA,* 246, 1223, 1981.
12. **Temin, P.**, *Taking Your Medicine: Drug Regulation in the United States,* Harvard University Press, Cambridge, Mass. and London, 1980, 42.
13. **Temin, P.**, *Taking Your Medicine: Drug Regulation in the United States,* Harvard University Press, Cambridge, Mass. and London, 1980, 47.
14. Good Laboratory Practices, *Federal Register,* 43, 60013, 1978.
15. Good Manufacturing Practices, *Federal Register,* 43, 45007, 1978.

INDEX

Printed and bound by CPI Group (UK) Ltd, Croydon, CR0 4YY

22/10/2024

01777630-0001